Evolutionary Essays: A Thermodynamic Interpretation of the Evolution

Evolutionary Essays

A Thermodynamic Interpretation of the Evolution

Sven E. Jørgensen

Professor, Dr. Eng., Dr. Scient.,
Dr. Hon. Causa.
Copenhagen University
University Park 2

ELSEVIER

Amsterdam • Boston • Heidelberg • London • New York • Oxford
Paris • San Diego • San Francisco • Singapore • Sydney • Tokyo

Elsevier
Radarweg 29, PO Box 211, 1000 AE Amsterdam, The Netherlands
The Boulevard, Langford Lane, Kidlington, Oxford OX5 1GB, UK

First edition 2008

British Library Cataloguing in Publication Data
A catalogue record for this book is available from the British Library

Library of Congress Cataloging-in-Publication Data
A catalog record for this book is available from the Library of Congress

ISBN: 978-0-444-52996-1

For information on all Elsevier publications
visit our website at books.elsevier.com

Printed and bound in Great Britain

08 09 10 10 9 8 7 6 5 4 3 2 1

CONTENTS

PREFACE

THERMODYNAMICS' HARMONY

"So one thing shall never cease to arise
out of another and life is granted to none
or freehold, to all as tenants on lease."

Lucretius, *De rerum natura, III, 970–971.*

If Tycho Brahe was a dragon guarding a treasure (his astronomical observations), the young mathematician and author of astrological almanacs, Johannes Kepler, was a modest popular antihero who stole the treasure and used it to forge the new astronomy through mathematical algorithms. Thus Kitty Ferguson describes the strange pair, Brahe and Kepler, in her essay "The Nobleman and his Housedog" (2002).

As far as the thermodynamics of life are concerned, the eminent Danish scientist Sven Erik Jørgensen now holds the treasure of the "man with the golden nose" (Tycho Brahe) through long years of work and scientific observation of ecosystems, refined mathematical skill and a talent for creating new equations and models to explain the theory of ecological systems.

It seems a strange coincidence that Kitty Ferguson was a musician and played in an orchestra conducted by Igor Stravinsky while Jørgensen made of this interesting book a "fugue by Bach", following an idea of Yuri Svirezhev.

Brahe, Kepler and Jørgensen share an idea of the world and of nature that can be described as *harmony*. The combination of mathematics and music is not new. Examples appear in the famous essay of Douglas Hofstadter, "Gödel, Escher, Bach: An Eternal Golden Braid". The theorem of Gödel is a recurring theme in Jørgensen's work.

Sven Jørgensen has often underlined three statements to describe modern physics:

1. Everything is relative (Einstein).
2. Everything is uncertain (Niels Bohr, Werner Heisenberg and Erwin Schrödinger).
3. Everything is irreversible (Prigogine).

Since Thermodynamics is the real deep core of physics as far as evolution and life are concerned, these evolutionary essays pave a fundamental road to understand ecosystems and living patterns.

According to Jørgensen, eco-exergy is a fundamental concept to describe evolutionary processes: "all real processes are irreversible which implies that exergy inevitably is lost. Exergy is not conserved". Eco-exergy is also the bridge between equilibrium and far from equilibrium (living) systems. Jørgensen is relating these views with the concept of dissipative structure by Prigogine.

Thermodynamic models are applied to different organisms, from *Archaea* to *Homo Sapiens*. The variety of examples is astonishing: from chemical reactions (ATP, oxidation, photosynthesis, biomasses) to growth forms, from the role of solar radiation to new animals emergent (in terms of evolutionary index), from the role of information to the epigenetic inheritance, including symbolic and semantic arguments.

The chapter on cosmic evolution is particularly interesting, though not in line with my personal point of view. I remember talking about this problem late into the night with Ilya Prigogine. In this chapter, Miller's experiments are discussed in terms of thermodynamics. New protagonists appear on the scene: eukaryote cells, pluricellular organization and the Cambrian Explosion, up to modern times.

The third movement is a holistic, thermodynamic interpretation of the evolution: let me underline the chapter dedicated to the evolution of diversity. Biodiversity is the main characteristic of life and is strictly related to eco-exergy, to the negentropic role and to ascendency (a measure of how far from equilibrium is a system). Many figures and models are presented. The intriguing part dedicated to the evolution of networks and to the emergence of novelties is illustrated with many examples, in terms of *eco-exergy flows* and *eco-exergy ascendency*. These thermodynamic functions give a clear idea of evolution, whereas evolution cannot be explained in terms of mere conservative properties, as mass and energy. In this context an important conclusion by Jørgensen is that "the origin of life is older than the Earth".

The eco-exergy embodied in the organisms has increased 400-fold, but the ability of the organisms to utilize eco-exergy—Jørgensen asserts—has increased 450,000 fold! "The eco-exergy flow represents the rate at which organisms can influence and utilize their environment".

All the important reations between thermodynamics and ecosystems, thermodynamics and evolution, thermodynamics and the chemistry of living species are treated in this book with deep and detailed analysis, offering to the reader an encyclopaedia of the matter, full of examples, calculations, data, rich in references. Moreover, all this is carried on by an evolutionary approach using the musical movements as a sort of evolutionary steps.

The book is an "attempt to capture the evolution as a continous irreversible process, that can be interpreted by use of thermodynamics. The three growth forms—growth and development of biomass, networks and information—are playing an important role in this interpretation."

Thermodynamics is the only branch of physics and chemistry able to understand open and evolutionary systems. Thermodynamics stays to physics as logic stays to philosophy. In the mean time Jørgensen is asserting, correctly, that "a complete, detailed, comprehensive and consistent description of an open system can never be obtained. Only a partial, though useful description (model) covering one or a few out of many views can be achieved", since "ecosystems are irreducible and due to their enormous complexity which prohibits us to know all details, we will only be able to indicate the propensities of their development. Ecosystems are not deterministic systems".

The above statements are in line with Gödel's theorem and with the Prigogine's ideas of the creation of order out of chaos and of self-organization.

The first two statements at the beginning of Section 2.7 are a piece of poetry: *the poem of the emergence of novelties*.

This fundamental book provides a *Table of Elements of Evolution* in terms of thermodynamic functions and processes. I think that the contributions of Ilya Prigogine, Robert Ulanowicz and Sven Erik Jørgensen now give scientists a basis for endeavouring to understand life, emergence of novelty, and to move along this musical path.

Enzo Tiezzi
Siena, 1 April 2008

PRELUDE

Escaping far from Hell, would we approach to Paradise?
Agrippa from Pisa

ב) והנה בהיות אור הא״ס נמשך.
בבחינת (ה) קו ישר תוך החלל
הנ״ל. לא נמשך ונתפשט (ו) תיכף
עד למטה, אמנם היה מתפשט לאט
לאט. רצוני לומר, כי בתחילה התי-
חיל קו האור להתפשט. ושם תיכף
(ז) בתחילת התפשטותו בסוד קו
נתפשט ונמשך ונעשה. כעין (ח)
גלגל אחד עגול מסביב.

"Multaque tum interisse animantum saecla necesses
Nec potuisse propagando procudere prolem.
Nam quaecumque vides vesci vitalibus auris,
Aut dolus, aut vitrus, aut denique mobilitas est
Ex ineunte aevo genus id tutata reservans"

Lucretius **De Rerum Natura.** Liber 5, 855

- A natural continuation of **"Toward a Thermodynamic Theory of Ecological Systems"**.

I planned to write this book about evolutionary thermodynamic together with Yuri Svirezhev, three years ago, after we have successfully finished the book *Toward a Thermodynamic Theory of Ecological Systems*. It was Yuri's idea to organize the book as a fugue by Bach—a composer that we both loved and often enjoyed. My dear friend Yuri passed away February 2007. We had planned to have brainstorming meetings during the spring and summer 2007. He had probably several splendid mathematical ideas about the content of the book, but nothing concrete was written down in a form that could be used by me. I had, therefore, to finish the book without the help from Yuri during the second half of year 2007. With Yuri, I would have written another book—more mathematically. When we wrote *Toward a Thermodynamic Theory of Ecological Systems*, he used to call me once a week and comment on what I have written and sent to him: "you are absolutely right, Sven, but you need the elegancy of mathematics". I hope, however, that the book in spite of the lack of mathematical elegancy presents a new, thermodynamic and holistic view of the evolution, which can open for a wide discussion on this core topic in biclogy and ecology. I have followed Yuri's proposal to use a fugue as the template for the book. Dux is the evolution and Comes the thermodynamic.

I would like to dedicate the book to the memory of Yuri Svirezhev, who was a great scientist and a very warm and open-minded person.

Sven E. Jørgensen Copenhagen

First Movement

Thermodynamics, Ecosystems and the Evolutionary Mechanisms

1

An introduction to thermodynamics and information theory

1.1 INTRODUCTION

The thermodynamic interpretation of ecosystems and their functions, presented and applied in the book *Towards a Thermodynamic Theory for Ecological Systems*, by S.E. Jørgensen and Y. Svirezhev, is expanded in this book and applied on the interpretation of the evolution. For the readers who have not read *Towards a Thermodynamic Theory for Ecological Systems*, it is therefore necessary to review the basic thermodynamics that was applied to develop a thermodynamic theory for ecosystems. The applied thermodynamics will be reviewed very briefly in this chapter. The following two chapters will discuss the mechanisms of evolution and their thermodynamic interpretation. The chapters will demonstrate how our thermodynamic ecosystem theory can be turned in the direction of an evolutionary theory.

The second movement of this book presents the history of evolution step by step with a translation to thermodynamic by the use of the concept of eco-exergy. Each chapter in this movement focuses on a specific part of the evolution, known from fossils and the evolutionary biology.

The last movement of the book attempts to understand the entire evolution as one continous process—see so to say the forest and not the individual trees. Both eco-exergy, eco-exergy flow and ascendency are considered in this attempt to capture the evolution as a continous irreversible process, that can be interpreted by use of thermodynamics. The three growth forms—growth and development of biomass, networks and information—are playing an important role in this interpretation.

1.2 THERMODYNAMICS AND ECOSYSTEMS

The basic thermodynamic laws are of course also valid for ecosystems. It means that the conservation law is valid, and as nuclear processes of very minor importance in ecosystems, both energy and mass are conserved.

Ecosystems are open systems in the sense that they are open for mass and energy transfers. Ecosystems receive energy from solar radiation and receive water from precipitation, dry deposition from the atmosphere, inputs by wind and inflows and outflows of various components plus emigration or immigration of species. A system that is closed to inputs and outputs of energy and mass is called an isolated system, while a system that is closed to inputs and outputs of mass, but open to energy transfers is denoted a closed system. A non-isolated system is a closed or open system. If an ecosystem was isolated, they would inevitably move towards the thermodynamic equilibrium and become a dead system with no gradients to do work—or as expressed in the following equations $dG = 0$ and $dS = 0$ at a maximum S value and a minimum G value, where G expresses Gibb's free energy and S the entropy. The openness explains why an ecosystem can maintain life and stay far from thermodynamic equilibrium, because maintenance of life requires input of energy which of course only is possible if an ecosystem is at least non-isolated.

The use of the second law of thermodynamics for open systems is crucial. At the first glance, it looks like ecosystems violate the second law of thermodynamics because they are moving away from thermodynamic equilibrium by the formation of a biological structure. Ecosystems receive, however, energy in the form of solar radiation, which can compensate for the steady transfer of work to heat. Several proposals on how to apply the second law of thermodynamics on systems far from thermodynamic equilibrium have been given, but before we turn to this central issue for open systems, it will be discussed what openness implies for the properties of ecosystems.

1.3 PHYSICAL OPENNESS

An energy balance equation for ecosystems might be written as follows in accordance with the principle of energy conservation:

$$E_{\text{cap}} = Q_{\text{evap}} + Q_{\text{resp}} + \cdots + \Delta E_{\text{bio}} \tag{1.1}$$

Here E_{cap} is the external energy captured per unit of time. Q_{evap} is the heat generated by evaporation, Q_{resp} is the heat generated by respiration and ΔE_{bio} is energy stored in the biological components – all per unit of time. A part of the incoming energy, solar radiation, being the main source for the ecosystems on Earth, is captured and a part is reflected unused, determining the albedo of the globe. The more biological structure an ecosystem possesses, the more of the incoming energy it is able to capture, that is the lower the albedo. The structure functions as an umbrella capturing the incoming solar radiation.

In ecosystem steady states, the formation of biological compounds (anabolism) is in approximate balance with their decomposition (catabolism). That is, in energy terms:

$$\Delta E_{\text{bio}} \approx 0, \quad \text{and} \quad E_{\text{cap}} \approx Q_{\text{evap}} + Q_{\text{resp}} + \cdots \tag{1.2}$$

The energy captured can in principle be any form of energy (electromagnetic, electrical, magnetic, chemical, kinetic, etc.) but for the ecosystems on Earth the short-wave energy of solar radiation (electromagnetic energy) plays the major role. The energy captured per unit of time is, however, according to Equation 1.2 used to pay the cost of maintenance per unit of time $= Q_{\text{evap}} + Q_{\text{resp}} \ldots$, The overall results of these processes require that E_{cap} be > 0, which entails openness (or at least non-isolation).

The following reaction chain summarizes the consequences of energy openness (Jørgensen et al., 1999): *source*: solar radiation → *anabolism* (charge phase): incorporation of high-quality energy, with entrained work capacity (and information), into complex biomolecular structures, entailing antientropic system movement away from equilibrium → *catabolism* (discharge phase): deterioration of structure involving release of chemical bond energy and its degradation to lower states of usefulness for work (heat) → *sink*: dissipation of degraded (low work capacity and high entropy) energy as heat to the environment (and, from Earth, to deep space), involving entropy generation and return towards thermodynamic equilibrium. Work capacity is expressed by $(g_s - g_o)Q$, where $(g_s - g_o)$ is a gradient in quality between the considered system, g_s, and a reference system, g_o, while Q is the corresponding quantity. For instance, for electrical energy g is the voltage and Q is the charge and for chemical energy g is the chemical potential and Q is the number of models.

This above-described chain can also be expressed in terms of matter: *source*: geochemical substrates relatively close to thermodynamic equilibrium → *anabolism*: inorganic chemicals are moulded into complex organic molecules (with low probability, it means that the equilibrium constant for the formation process is very low, low entropy, and high distance from thermodynamic) → *catabolism*: synthesized organic matter is ultimately decomposed into simple inorganic molecules again; the distance from thermodynamic equilibrium decreases, and entropy increases → *cycling*: the inorganic molecules, returned to near-equilibrium states, become available in the nearly closed material ecosphere of Earth for repetition of the matter charge–discharge cycle.

Input environments of ecosystems serve as sources of high-quality energy whose high contents of work and information and low entropy raise the organizational states of matter far from equilibrium. Output environments, in contrast, are sinks for energy and matter lower in work capacity, higher in entropy, and closer to equilibrium. Since, in the organization of ecosystems, output environments feed back to become portions of input environments, living systems operating in the ecosphere, which are energetically non-isolated but materially nearly closed, must seek an adaptive balance between these two aspects of their environmental relations in order to sustain their continued existence.

Following from Gödel's Theorem, a scientific description can only be given from outside open systems. Natural science cannot be applied to isolated systems (the universe is considered open due to the expansion) at all. A complete, detailed, comprehensive and consistent description of an open system can never be obtained. Only a partial, though useful description (model) covering one or a few out of many views can be achieved.

Due to the enormous complexity of ecosystems, we cannot know all the details of ecosystems. When we cannot know all the details, we are not able to describe fully the initial stage and the processes that determine the development of the ecosystems—as expressed above, ecosystems are therefore irreducible. Ecosystems are not deterministic because we cannot provide all the observations that are needed to give a full deterministic description; or as expressed by Tiezzi, ecosystems do play dice (Tiezzi, 2003, 2006).This implies that our description of ecosystem developments must be open to a wide spectrum of possibilities. It is consistent with the application of chaos and catastrophe theory; see, for instance, Jørgensen (1992a,b, 1995a,b, 2002). Ulanowicz (1997) makes a major issue of the necessity for systems to be causally open in order to be living—the open possibilities may create new pathways for development which may be crucial for survival and further evolution in a non-deterministic world. He goes so far as to contend that a mature insight into the evolutionary process is impossible without a revision of our contemporary notions on causality. Ulanowicz (1997) uses the concept of propensity to come around the problem of causality. On the one side, we are able to relate the development with the changing internal and external factors of ecosystems. On the other side, due to the uncertainty in our predictions of development caused by our lack of knowledge about all the details, we are not able to give deterministic descriptions of the development, but we can only indicate which propensities will be governing.

To conclude: Ecosystems are irreducible and due to their enormous complexity which prohibits us to know all details, we will only be able to indicate the propensities of their development. Ecosystems are not deterministic systems.

1.4 THE SECOND LAW OF THERMODYNAMICS INTERPRETED FOR OPEN SYSTEMS

Let us first expand on the conclusions that we already have made to give more detail on the differences between isolated and open systems and thereby understand better the application of the second law of thermodynamics on open systems (Jørgensen et al., 1999).

If ecosystems were isolated, no energy or matter could be exchanged across their boundaries. The systems would spontaneously degrade their initially contained work capacity, that is reduce the gradients and increase their entropy, corresponding to a loss of order and organization, and an increase in the randomness of their constituents and microstates. This dissipation process would cease at equilibrium, where no further motion or change would be possible. The physical manifestation would ultimately be a meltdown to the proverbial "inorganic soup" containing degradation products dispersed equi-probably through the entire volume of the system. Gradients of all kinds would be eliminated, and the system would be frozen in time in a stable, fixed configuration. The high-energy chemical compounds of biological systems, faced suddenly with isolation, would decompose spontaneously to compounds with high-entropy contents. The process would be progressive, to higher and higher entropy, and by presence of oxygen would end with a mixture of inorganic residues—carbon dioxide and water, nitrates, phosphates, sulphates and so on. These simpler compounds could never be reconfigured into the complex molecules necessary to carry on life processes without the input of new low-entropy energy to be employed in biosynthesis. An isolated ecosystem could therefore in the best case sustain life for only a limited period of time, less than that required from the onset of isolation to reach thermodynamic equilibrium. This local situation is comparable to the "heat death" of the universe, proposed by physicists a century ago as the ultimate outcome of the second law of thermodynamics. Thus, thermodynamic equilibrium is the global attractor for all physical processes isolated from their surroundings. Having reached it, no further changes are possible. In this "frozen" state even time would have no meaning as its passage could not be verified by reference to any changes.

Observations of properties could not be made, only inferred, because observation requires some kind of exchanges between the system and an observer. There would be no internal processes, because no gradients would exist to enable them. There would only be uninterrupted and uninterruptable stillness and sameness which would never change. The system would be completely static at the thermodynamic equilibrium. Thus, in a peculiar way, isolated systems can only be pure abstractions in reality, submitting neither to time passage, change, nor to actual observation. They are the first "black holes" of physics and the antithesis of our systems plus their environments which are the core model for systems ecology. No ecosystem could ever exist and be known to us as an isolated system.

The change in entropy for an *open* system, $\mathrm{d}S_{\text{system}}$, consists of an external, exogenous contribution from the environment, $\mathrm{de}S = S_{\text{in}} - S_{\text{out}}$, and an internal, endogenous contribution due to system state, $\mathrm{di}S$, which must always be positive by the second law of thermodynamics (Prigogine, 1980). Prigogine (1980) uses the concept of entropy and the second law of thermodynamics far from thermodynamic equilibrium, which is outside the framework of classical thermodynamics, but he uses the concepts only locally.

There are three possibilities for the entropy balance:

$$\mathrm{d}S_{\text{system}}/\mathrm{d}t = \mathrm{de}S/\mathrm{d}t + \mathrm{di}S/\mathrm{d}t > 0 \tag{1.3}$$

$$\mathrm{d}S_{\text{system}}/\mathrm{d}t = \mathrm{de}S/\mathrm{d}t + \mathrm{di}S/\mathrm{d}t < 0 \tag{1.4}$$

$$\mathrm{d}S_{\text{system}}/\mathrm{d}t = 0 \tag{1.5}$$

The system looses order in the first case. Gaining order (case 2, Equation 1.4) is *only* possible if $-\mathrm{de}S > \mathrm{di}S > 0$. This means that if order is to be created in a system ($\mathrm{d}S_{\text{system}} < 0$), $\mathrm{de}S$ must be < 0 and therefore $S_{\text{in}} < S_{\text{out}}$.

Creation of order in a system must be associated with a greater flux of entropy out of the system than into the system. This implies that the system must be open or at least non-isolated.

Case 3, Equation 1.5, corresponds to a stationary situation, for which Ebeling et al. (1990) use the following two equations for the energy (U) balance and the entropy (S) balance:

$$\mathrm{d}U/\mathrm{d}t = 0 \quad \text{or} \quad \mathrm{de}U/\mathrm{d}t = -\mathrm{di}U/\mathrm{d}t = 0 \tag{1.6}$$

and

$$\mathrm{d}S_{\text{system}}/\mathrm{d}t = 0 \quad \text{or} \quad \mathrm{de}S/\mathrm{d}t = -\mathrm{di}S/\mathrm{d}t = 0 \tag{1.7}$$

Usually the thermodynamic processes are isotherm and isobar. This implies that we can interpret the third case (Equations 3.19–3.21) by use of the free energy:

$$\mathrm{de}G/\mathrm{d}t = T\,\mathrm{di}S/\mathrm{d}t > 0$$

It means that a "status quo" situation for an ecosystem requires input of free energy to compensate for the loss of free energy and the corresponding formation of heat due to maintenance processes, that is respiration and evapotranspiration. If the system is not receiving a sufficient amount of free energy, the entropy will increase. If the entropy of the system will continue to increase, the system will approach thermodynamic equilibrium—the system will die. This is in accordance with Ostwald (1931, 1978): life without the input of free energy is not possible.

The entropy produced by the life processes can be exported by three processes (Svirezhev, 1990): (1) transfer of heat to the environment; (2) exchange of material with the environment; (3) biochemical processes in the system. The first process (heat transfer) is of particular importance.

An energy flow of about 10^{17} W by solar radiation ensures the maintenance of life on Earth. The surface temperature of the Sun is 5800 K and of the Earth in average about 280 K. This implies that the following export of entropy per unit of time takes place from the Earth to the open space:

$$10^{17}\,\mathrm{W}\left(\frac{1}{5800}\mathrm{K} - \frac{1}{280}\mathrm{K}\right) \approx 4.1014\,\mathrm{W/K} \tag{1.8}$$

corresponding to $1\,\mathrm{W/m^2\,K}$.

Ecosystems can maintain a certain concentration of low-entropy compounds against the second-law dissipation gradient, because they are not isolated. Ecosystems receive a continuous supply of free energy or negentropy (potential entropy, not yet released (see Schrödinger, 1944)) from outside to compensate for the positive entropy produced internally as a consequence of the second law of thermodynamics ($\mathrm{di}S > 0$). On Earth, solar radiation is the main source of this input of free energy, negentropy or low-entropy energy. The incoming energy has lower entropy, while the outgoing energy has higher entropy.

All ordered structures require low entropy for maintenance, and therefore for a system to maintain structure or increase its internal order it must receive input of low-entropy energy from external sources. *Structure*, in this context, is a spatial or temporal order describable in terms of information theory. Prigogine uses the term *dissipative structure* to denote self-organizing systems, thereby indicating that such systems dissipate energy (produce entropy) for the maintenance of their organization (order). The following conclusions are appropriate.

All systems, because they are subject to the second law of thermodynamics, are inherently dissipative structures. To offset the dissipative processes they require inputs of low-entropy energy to maintain or produce more internal organized structure, measurable in terms of information content. Thus, all real systems must be open or, at least, non-isolated.

As previously noted, ecosystems, in common with all real systems, have a global attractor state—thermodynamic equilibrium. Through their openness they avoid reaching this state by importing low entropy, or matter carrying information from their surroundings. This anabolism combats and compensates for the catabolic deterioration of structure; the two processes operate against one another. Note that the equilibrium "attractor" represents a resting or refractory state, one that is passively devolved to if system openness or non-isolation is compromised (Jørgensen et al., 1999).

1.5 THE THIRD LAW OF THERMODYNAMICS APPLIED ON OPEN SYSTEMS

The first law of thermodynamics is often applied on ecosystems, first of all when energy balances of ecosystems are made. Also the second law of thermodynamics is applied on ecosystems, when we consider the entropy production of ecosystems as a consequence of the maintenance of the systems far from thermodynamic equilibrium. This section is concerned with the application of the third law of thermodynamics on ecosystems.

The lesser-known third law of thermodynamics states that the entropies, S^0, of pure chemical compounds are zero, and that entropy production, ΔS^0, by chemical reactions between pure crystalline compounds is zero at absolute temperature, 0 K. The third law implies, since both $S^0 = 0$ (absolute order) and $\Delta S^0 = 0$ (no disorder generation), that disorder does not exist and cannot be created at absolute zero temperature. But at temperatures > 0 K disorder can exist ($S_{\text{system}} > 0$) and be generated ($\Delta S_{\text{system}} > 0$). The third law defines the relation between entropy production, ΔS_{system}, and the Kelvin temperature, T:

$$\Delta S_{\text{system}} = \int_0 T \Delta c_{\text{p}} \, \mathrm{d}\ln T + \Delta S^{\text{o}} \tag{1.9}$$

where Δcp is the increase in heat capacity by the chemical reaction. Since order is absolute at absolute zero, its further creation is precluded there. At higher temperatures, however, order can be created.

Entropy production implies that degradation of energy from a state of high utility (large T) to a state of low utility (small T) occurs; compare also with Carnot's Cycle. Ecosystems have in other words a global attractor state, the thermodynamic equilibrium, but will never reach this state as long as they are not isolated and receives exergy from outside to combat the decomposition of its compounds. As ecosystems *have* an energy through-flow, the attractor becomes the steady state, where the formation of new biological compounds is in balance with the decomposition processes (see Equation 1.2). As seen from these perspectives of the second law of thermodynamics for open (non-isolated) systems, it is vital to ecosystem to be non-isolated.

It has been stated a few times that it is *necessary* for an ecosystem to transfer the generated heat (entropy) to the environment and to receive low-entropy energy (solar radiation) from the environment for the formation of dissipative structure. The next obvious question would be: will energy source and sink also be *sufficient* to initiate the formation of dissipative structure, which can be used as the source for entropy combating processes?

The answer to this question is "yes". It can be shown by the use of simple model systems and basic thermodynamics; see Morowitz (1968, 1978, 1992). He shows that a flow of energy from sources to sinks leads to an internal organization of the system and to the establishment of element cycles. The type of organization is, of course, dependent on a number of factors: the temperature, the elements present, the initial conditions of the system and the time available for the development of organization. It is characteristic for the system, as pointed out above, that the steady state of an open system does *not* involve chemical equilibrium.

Prigogine and his colleagues have shown that open systems that are exposed to an energy through-flow exhibit coherent self-organization behaviour and are known as dissipative structures. Formations of complex organic compounds from inorganic matter as mentioned above are typical examples of self-organization. Such systems can remain in their organized state by exporting entropy outside the system, but are dependent on outside energy fluxes to maintain their organization, as was already mentioned and emphasized above. Glansdorff and Prigogine (1971) have shown that the thermodynamic relationship of far from equilibrium dissipative structures is best represented by coupled non-linear relationships, that is autocatalytic positive feedback cycles.

Creation of order in open systems with through-flow of energy is inevitable. On Earth, the surface temperature difference between the Sun and the planet guarantees this. Morowitz (1968, 1992) showed that energy through-flow is sufficient to produce cycling, a prerequisite for the ordering processes characteristic of living systems.

A system at 0 K on the other hand is without any creative potential, because no dissipation of energy can take place at this temperature. A temperature greater than 2.726 ± 0.01 K, where 2.726 K is the temperature of deep space, is therefore required before order can be created. At 0 K the world is dead and still, as the temperature is a measure of the velocity of atoms. The so-called Bose–Einstein condensate is formed, where all atoms are the same and behave like one single atom. This was predicted by Bose and Einstein in the 1920s, but has recently been shown experimentally at temperatures very close to 0 K.

The velocity is zero at 0 K by definition and therefore determined without uncertainty. This explains that the position is undetermined according to Heisenberg's uncertainty equation. At 0 K there is therefore no structure, no gradients and no complexity. No entropy can be formed because all mass is everywhere and nowhere and without form and structure. There is no disorder to create and therefore no entropy to produce. The system is trapped between complete order because all the mass occupies all the space, and complete disorder because all the space is occupied by the mass—a complete dissipation has taken place. At 0 K, no creativity is possible, no differences (gradients), no structure and even no physical activity, because all velocities are zero. Everything is dull and dead. Even the light has stopped. Time has no meaning, because time is determined by the rate of changes.

These extreme conditions at 0 K elucidate the meaning of the concept "entropy". Entropy is the price we have to pay for order, structure, organization and creativity, and without entropy production, there would be no order, structure, organization and creativity.

Moreover, it explains the meaning behind the second law of thermodynamics. Because heat is an energy form which is generated by transformation of all other energy forms and because a 100% effective transformation of heat to work cannot take place, according to the Carnot equation (the cold reservoir can never be maintained at 0 K): energy that can do work is inexorably lost to energy that cannot do work. This is the condition which is imposed on us: time (and all reactions) are irreversible.

1.6 WHAT IS EXERGY?

Previously in this chapter, we have used the term "work capacity" to express the ability of a part of the total energy to perform work in contrast to the heat energy at the temperature of the environment, that is without work capacity. The classical thermodynamics is using the G-function to cover the work capacity; but when we are dealing with very far from thermodynamic systems, we cannot any longer use state variables that are independent of the pathway. Furthermore, we need in different situations different reference states. Therefore, we have to define a work capacity that can be used also for very far from thermodynamic equilibrium systems (Szargut et al., 1988).

Exergy is defined as the amount of work (=entropy-free energy) a system can perform when it is brought into thermodynamic equilibrium with its environment (Jørgensen et al., 1999). Figure 1.1 illustrates the definition. The considered system is characterized by the extensive state variables S, U, V, N_1, N_2, N_3 ...., where S is the entropy, U is the energy, V is the volume and N_1, N_2, N_3 .... are moles of various

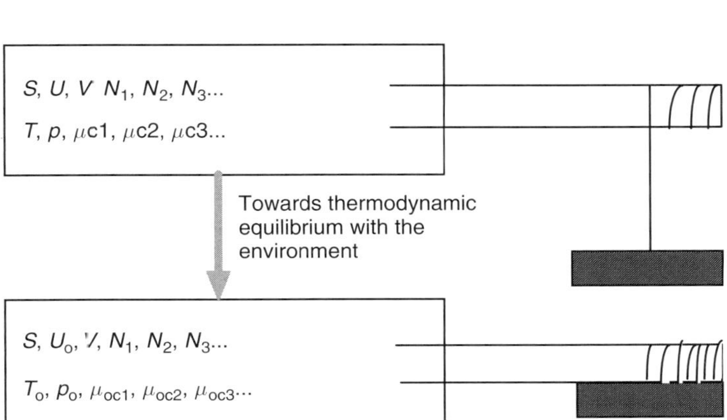

Figure 1.1 Definition of exergy is shown.

chemical compounds, and by the intensive state variables, T, p, μ_{c1}, μ_{c2}, $\mu_{c3}\ldots$. The system is coupled to a reservoir, a reference state, by a shaft. The system and the reservoir are forming a closed system. The reservoir (the environment) is characterized by the intensive state variables T_o, p_o, μ_{c1o}, μ_{c2o}, μ_{c3o} and as the system is small compared with the reservoir, the intensive state variables of the reservoir will not be changed by interactions between the system and the reservoir. The system develops towards thermodynamic equilibrium with the reservoir and is simultaneously able to release entropy-free energy to the reservoir. During this process the volume of the system is constant as the entropy-free energy must be transferred through the shaft only.

The entropy is also constant as the process is an entropy-free energy transfer from the system to the reservoir, but the intensive state variables of the system become equal to the values for the reservoir. The total transfer of entropy-free energy in this case is the exergy of the system. It is seen from this definition that exergy is dependent on the state of the total system (=system + reservoir) and not dependent entirely on the state of the system. Exergy is therefore not a state variable. In accordance with the first law of thermodynamics, the increase of energy in the reservoir, ΔU, is

$$\Delta U = U - U_0 \tag{1.10}$$

where U_o is the energy content of the system after the transfer of work to the reservoir has taken place. According to the definition of exergy, Ex, we have

$$Ex = \Delta U = U - U_0$$

As

$$U = TS - pV + \sum_c \mu_c N_i \tag{1.11}$$

(when we consider only heat, spatial energy (displacement work) and chemical energy, see any textbook in thermodynamics), and correspondingly for U_o:

$$U_o = T_o S - p_o V + \sum_c \mu_{co} N_i \tag{1.12}$$

we get the following expression for exergy, excluding of course in this case kinetic energy, potential energy, electrical energy, radiation energy and magnetic energy:

$$Ex = S(T - T_o) - V(p - p_o) + \sum_c (\mu_c - \mu_{co}) N_i \tag{1.13}$$

Notice that the above-shown equation also emphasizes that exergy is dependent on the state of the environment (the reservoir = the reference state), as the exergy of the system is dependent on the intensive state variables of the reservoir. Notice furthermore that exergy is not conserved—only if entropy-free energy is transferred, which implies that the transfer is reversible. All processes in reality are, however, irreversible, which means that exergy is lost (and entropy is produced). Loss of exergy and production of entropy are two different descriptions of the same reality, namely that all processes are irreversible, and we unfortunately always have some loss of energy forms which can do work to energy forms which cannot do work (heat at the temperature of the environment) (see also Jørgensen, 2002). So, the formulation of the second law of thermodynamics by use of exergy is

All real processes are irreversible which implies that exergy inevitably is lost. Exergy is not conserved,

while energy of course is conserved by all processes according to the first law of thermodynamics. It is therefore wrong to discuss, an energy efficiency of an energy transfer, because it will always be 100%, while the exergy efficiency is of interest, because it will express the ratio of useful energy to total energy which always is less than 100% for real processes. All transfers of energy imply that exergy is lost because energy is transformed to heat at the temperature of the environment.

It is therefore of interest to set up for all environmental systems an exergy balance in addition to an energy balance. Our concern is loss of exergy, because that means that "first class energy" which can do work is lost as "second class energy" (heat at the temperature of the environment) which cannot do work. So, the particular properties of heat and that temperature is a measure of the movement of molecules, given limitations in our possibilities to utilize energy to do work. Due to these limitations, we have to distinguish between exergy which can do work and anergy which cannot do work, and all real processes imply inevitably a loss of exergy as anergy (see also next section).

Exergy seems more useful to apply than entropy to describe the irreversibility of real processes, as it has the same unit as energy and is an energy form, while the definition of entropy is more difficult to relate to the concepts associated to our usual description of the reality. In addition, entropy is not clearly defined for "far from thermodynamic equilibrium systems", particularly for living systems (see, for instance, Tiezzi, 2003). Moreover, it should be mentioned that the self-organizing abilities of systems are strongly dependent on the temperature, as it is discussed in Jørgensen et al. (1999). Exergy takes the temperature into consideration as the definition shows, while entropy doesn't. It implies that exergy at 0 K is 0 and at minimum. The negative entropy is not expressing the ability of the system to do work (we may call it "the creativity" of the system as creativity requires work), but exergy becomes a good measure of "the creativity", which is increasing proportional with the temperature. Furthermore, exergy facilitates the differentiation between low-entropy energy and high-entropy energy, as exergy is entropy-free energy.

Finally, notice that information contains exergy. Boltzmann (1905) showed that the free energy of the information that we actually possess (in contrast to the information we need to describe the system) is $k \times T \times \ln I$, where I is the information we have about the state of the system, for instance, that the configuration is 1 out of W possible (i.e., that

$W = I$) and k is Boltzmann's constant $= 1.3803 \times 10^{-23}$ (J/molecules $\times$ deg). It implies that one bit of information has the exergy equal to $kT \ln 2$. Transformation of information from one system to another is often almost an entropy-free energy transfer. If the two systems have different temperatures, the entropy lost by one system is not equal to the entropy gained by the other system, while the exergy lost by the first system is equal to the exergy transferred and equal to the exergy gained by the other system, provided that the transformation is not accompanied by any loss of exergy. In this case, it is obviously more convenient to apply exergy than entropy.

Exergy of the system measures the contrast—it is the difference in free energy if there is no difference in pressure and temperature, as may be assumed for an ecosystem or an environmental system and its environment—against the surrounding environment. If the system is in equilibrium with the surrounding environment ,the technological exergy is of course zero. The only way to move systems away from equilibrium is to perform work on them. Therefore, it is reasonable to use the available work, that is the exergy, as a measure of the distance from thermodynamic equilibrium. As we know that ecosystems due to the through-flow of energy have the tendency to move away from thermodynamic equilibrium, loosing entropy or gaining exergy and information, we can put forward the following proposition of relevance for ecosystems.

Ecosystems attempt to develop towards a higher level of exergy

This description of exergy development in ecosystems makes it pertinent to assess the exergy of ecosystems. It is not possible to measure exergy directly—but it is possible to compute. If we presume a reference environment that represents the same system (ecosystem) at thermodynamic equilibrium, which means that all the components are inorganic at the highest possible oxidation state if sufficient oxygen is present (as much free energy as possible is utilized to do work) and homogeneously distributed in the system (no gradients), the situation illustrated in Figure 1.2 is valid. As the chemical energy embodied in the organic components and the biological structure contribute far most to the exergy content of the system, there seems to be no reason to assume a (minor) temperature and pressure difference between the system and the reference environment. Under these circumstances we can calculate the exergy, which we will name eco-exergy to distinguish from the technological exergy defined above, as coming entirely from the chemical energy:

$$\sum_{c} (\mu_c - \mu_{co}) N_i \tag{1.14}$$

This represents the non-flow chemical exergy. It is determined by the difference in chemical potential $(\mu_c - \mu_{co})$ between the ecosystem and the same system at thermodynamic equilibrium. This difference is determined by the concentrations of the considered components in the system and in the reference state (thermodynamic equilibrium), as it is the case for all chemical processes. We can measure the concentrations in the ecosystem, but the concentrations in the reference state (thermodynamic

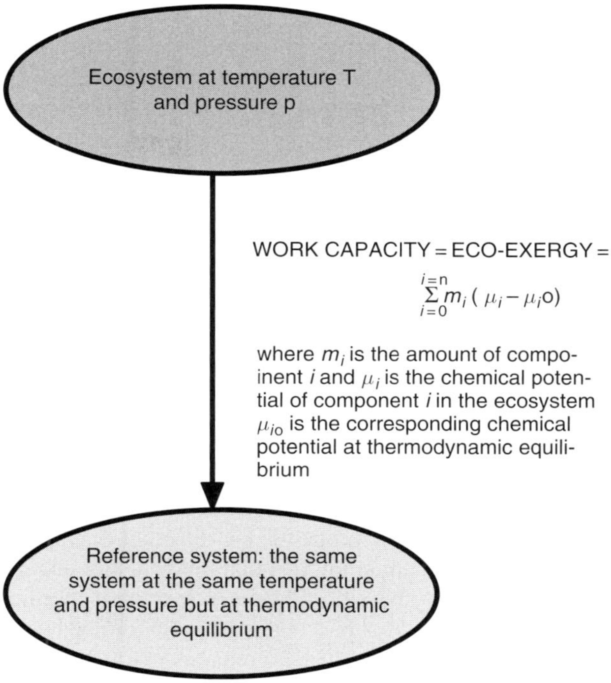

Figure 1.2 The exergy content of the system is calculated in the text for the system relative to a reference environment of the same system at the same temperature and pressure, but as an inorganic soup with no life, biological structure, information or organic molecules.

equilibrium) can be based on the usual use of chemical equilibrium constants. If we have the process:

$$\text{Component A} \leftrightarrow \text{inorganic decomposition products} \tag{1.15}$$

It has a chemical equilibrium constant, K:

$$\text{K} = [\text{inorganic decomposition products}]/[\text{Component A}] \tag{1.16}$$

The concentration of component A at thermodynamic equilibrium is difficult to find, but we can find it from the probability of forming A from the inorganic components.

Eco-exergy is a concept close to Gibb's free energy (De Wit, 2005); but opposite to Gibb's free energy, eco-exergy has a different reference state from case to case (from ecosystem to ecosystem) and it can furthermore be used far from thermodynamic equilibrium, while Gibb's free energy in accordance to its exact thermodynamic definition is only a state function close to thermodynamic equilibrium.

We find by these calculations the eco-exergy of the system compared with the same system at the same temperature and pressure but in the form of an inorganic soup without any life, biological structure, information or organic molecules. As $\mu_c - \mu_{co}$ can be found from the definition of the chemical potential replacing activities by concentrations, we get the following expressions for the exergy:

$$Ex = RT \sum_{i=0}^{i=n} C_i \ln C_i/C_{i,\mathrm{o}} \tag{1.17}$$

where R is the gas constant (8.317 J/K moles = 0.08207 l atm/K moles), T is the temperature of the system, while C_i is the concentration of the ith component expressed in a suitable unit, for example for phytoplankton in a lake, C_i could be expressed as mg/l or as mg/l of a focal nutrient. $C_{i,\mathrm{o}}$ is the concentration of the ith component at thermodynamic equilibrium and n is the number of components. $C_{i,\mathrm{o}}$ is of course a very small concentration (except for $i = 0$, which is considered to cover the inorganic compounds), corresponding to a very low probability of forming complex organic compounds spontaneously in an inorganic soup at thermodynamic equilibrium. $C_{i,\mathrm{o}}$ is even lower for the various organisms, because the probability of forming the organisms is very low with their embodied information which implies that the genetic code should be correct.

By using this particular exergy based on the same system at thermodynamic equilibrium as reference, the eco-exergy becomes dependent only on the chemical potential of the numerous biochemical components that are characteristic for life. It is consistent with Boltzmann's statement that life is struggle for free energy.

The total eco-exergy of an ecosystem *cannot* be calculated exactly, as we cannot measure the concentrations of *all* the components or determine all possible contributions to the eco-exergy of an ecosystem. If we calculate the eco-exergy of a fox, for instance, the above-shown calculations will only give the contributions coming from the biomass and the information embodied in the genes, but what is the contribution from the blood pressure, the sexual hormones and so on? These properties are at least partially covered by the genes, but is that the entire story? We can calculate the contributions from the dominant components, for instance, by the use of a model or measurements that covers the most essential components for a focal problem. The *difference* in eco-exergy by *comparison* of two different possible structures (species composition) is decisive here. Moreover, eco-exergy computations give always only relative values, as the eco-exergy is calculated relatively to the reference system.

Notice that the definition of eco-exergy is very close to free energy. Eco-exergy is, however, a *difference* in free energy between the system and the same system at thermodynamic equilibrium. The reference system used is different for every ecosystem according to the definition of eco-exergy. In addition, free energy is not a state function far from thermodynamic equilibrium. Consider, for instance, the immediate loss of free energy (or let us use the term eco-exergy as already proposed to make the use of the concepts more clear) when an organism dies. A microsecond before the death the information can be used and after the death the information is worthless and should therefore not be included in the calculation of eco-exergy.

1.7 ECO-EXERGY AND INFORMATION

Information means "acquired knowledge". The thermodynamic concept of exergy and eco-exergy is closely related to information. A high local concentration of a chemical compound, for instance, with a biochemical function that is rare elsewhere, carries exergy and information. The eco-exergy can be calculated as the exergy lost by dispersion similar to Equation 1.17—only the sign is of course opposite. On the more complex levels, information may still be strongly related to eco-exergy but in more indirect ways. Information is also a convenient measure of physical structure. A certain structure is chosen out of all possible structures and defined within certain tolerance margins (Thoma, 1977).

Biological structures maintain and reproduce themselves by transforming energy and information, from one form to another. An important feature by life is that the information laid down in the genetic material is developed and transferred from one generation to the next. When biological materials are used to the benefit of mankind, it is in fact the organic structures and the information contained therein that are of advantage, for instance when using wood.

In statistical mechanics, entropy is related to probability. A system can be characterized by averaging ensembles of microscopic states to yield the macrostate. If W is the number of microstates that will yield one particular macrostate, the probability P that this particular macrostate will occur as opposed to all other possible macrostates is proportional to W. It can further be shown that

$$S = k \times \ln W \tag{1.18}$$

where k is Boltzmann's constant, 1.3803×10^{-23} J/(molecules·K). The entropy is a logarithmic function of W and thus measures the total number of ways that a particular macrostate can be constituted microscopically. $kAv = R$, where Av is Avogadro's number. Notice that entropy measures the information that we need.

S may be called the thermodynamic information, meaning the amount of information *needed* to describe the system, which must *not* be interpreted as the information that we actually possess. The information that we have becomes equal to negentropy ($-S$), which from a strict thermodynamic point of view does not exist. According to the third law of thermodynamics, entropy cannot be negative, and it would be more correct to use the term the free energy (or exergy) of the information we have $= kT \ln W$. Boltzmann also used the expression the free energy of the information (Boltzmann, 1905). The more microstates there are and the more disordered they are, the more information is required and the more difficult it will be to give a complete description of the system—and the higher is the entropy. It means that entropy + information = constant. When we gain information, entropy decreases, and when we loose information, entropy increases. The constant expresses the total information needed to know all the details of a considered system. When we have the full information about the considered system, information = constant and entropy = 0. When we have no knowledge at all about the system, the entropy = constant and information = 0.

If we consider that this book contains 10^6 signs selected among 40 possible signs, the entropy according to Equation $1.18 = 1.3803 \times 10^{-23}\, 10^6 \ln 40 = 5.09\, 10^{-17}$ J/K. The entropy is the disorder resulting from 1,000,000 signs selected among 40 possible types. If we, however, understand the information embodied in the 1,000,000 signs, that is we can read English, the free energy (exergy) of the information in this book at room temperature (about 300 K) becomes 1.527×10^{-14} J.

It is possible to distinguish between the exergy of information and of biomass (Svirezhev, 1998). p_i is defined as c_i/A, where

$$A = \sum_{i=1}^{n} C_i$$

is the total amount of matter in the system, is introduced as new variable in Equation 1.19:

$$Ex = A\,RT \sum_{i=1}^{n} pi \ln p_i/p_{io} + A \ln A/A_o \tag{1.19}$$

As $A \approx A_o$, exergy becomes a product of the total biomass A (multiplied by RT) and Kullback measure:

$$K = \sum_{i=1}^{n} p_i \ln(p_i/p_{io}) \tag{1.20}$$

where p_i and p_{io} are probability distributions, a posteriori and a priori to an observation of the molecular detail of the system. It means that K expresses the amount of information that is gained as a result of the observations. If we observe a system, which consists of two connected chambers, we expect the molecules to be equally distributed in the two chambers, that is $p_1 = p_2$ is equal to 1/2. If we, on the other hand, observe that all the molecules are in one chamber, we get $p_1 = 1$ and $p_2 = 0$.

Specific exergy is exergy relative to the biomass and for the ith component: Sp. ex. $I = Ex_i/c_i$. It implies that the total specific exergy per unit of area or per unit of volume of the ecosystem is equal to RTK.

Due to the incoming energy of solar radiation, an ecosystem is able to move away from the thermodynamic equilibrium—that is, the system evolves, obtains more information and organization.

The ecosystem must produce entropy for maintenance, but the low-entropy energy flowing through the system may be able to more than cover this production of disorder, resulting in an increased order or information of the ecosystem.

1.8 DISSIPATIVE STRUCTURE AND EXERGY

As an ecosystem is non-isolated, the entropy changes during a time interval, dt, can be decomposed into the entropy flux due to exchanges with the environment and the entropy production due to the irreversible processes inside the system such as diffusion, heat conduction and chemical reactions; see Section 1.3. The entropy Equations (1.3–1.5) can be expressed by use of exergy:

$$Ex/\mathrm{d}t = \mathrm{de}Ex/\mathrm{d}t + \mathrm{di}Ex/\mathrm{d}t \tag{1.21}$$

where deEx/dt represents the exergy input to the system and diEx/dt is the exergy consumed (is negative) by the system for maintenance and so on.

Equation 1.21 shows among other things that systems can maintain only a non-equilibrium steady state by compensating the internal exergy consumption with a positive exergy influx (deEx/d$t > 0$). Such an influx induces order into the system. In ecosystems the ultimate exergy influx comes from solar radiation, and the order induced is, for example biochemical molecular order.

If de$Ex > -$diEx (the exergy consumption in the system), the system has surplus exergy input, which may be utilized to construct further order in the system, or as Prigogine (1980) calls it: dissipative structure. The system will thereby move further away from the thermodynamic equilibrium. The evolution shows that this situation has been valid for the ecosphere on a long-term basis. In spring and summer ecosystems are in the typical situation that deEx exceeds $-$diEx. If de$Ex < -$diEx, the system cannot maintain the order already achieved, but will move closer to the thermodynamic equilibrium, that is it will loose order. This may be the situation for ecosystems during fall and winter or due to environmental disturbances.

1.9 HOW TO CALCULATE EXERGY OF ORGANIC MATTER AND ORGANISMS?

The following expression for the exergy per unit of volume has been presented; see Equation 1.17:

$$Ex = RT\sum_{i=0}^{i=n} c_i \ln c_i/c_{io} \qquad [ML^{-1}T^{-2}] \tag{1.22}$$

where R is the gas constant, T is the temperature of the environment, while c_i is the concentration of the ith component expressed in a suitable unit, for example for phytoplankton in a lake c_i could be expressed as mg/l or as mg/l of a focal nutrient. c_{io} is the concentration of the ith component at thermodynamic equilibrium and n is the number of components. c_{io} is very low for living component because the probability that living components are formed at thermodynamic equilibrium is very low. It implies that living components get a high exergy. c_{io} is not zero for organisms, but will correspond to a very low probability of forming complex organic compounds spontaneously in an inorganic soup at thermodynamic equilibrium. c_{io} on the other hand is high for inorganic components, and although c_{io} is still low for detritus, it is much higher than for living components.

Shieh and Fan (1982) have suggested to estimate the exergy of structurally complicated material on the basis of the elementary composition. This has, however, the disadvantage that a higher organism and a microorganism with the same elementary composition will get the same exergy which is in complete disagreement with the lower probability to form a more complex organism, that is the lower concentration of c_{io} in Equation 1.22.

The problem related to the assessment of c_{io} has been discussed and a possible solution proposed in Jørgensen et al. (1995). For dead organic matter, detritus, which is given the index 1, it can be found from classical thermodynamics to be approximately 18.7 kJ/g in average.

For the biological components, 2, 3, 4, …, N, the probability, p_{io}, consists at least of the probability of producing the organic matter (detritus), that is p_{1o}, and the probability, $p_{i,a}$, to find the correct composition of the enzymes determining the biochemical processes in the organisms. Living organisms use 20 different amino acids and each gene determines in average the sequence of about 700 amino acids (Li and Grauer, 1991). $p_{i,a}$, can be found from the number of permutations among which the characteristic amino acid sequence for the considered organism has been selected. It means that

$$p_{i,a} = a^{-Ng_i}[-] \tag{1.23}$$

where a is the number of possible amino acids $= 20$, N is the number of amino acids determined by one gene $= 700$ and gi is the number of non-nonsense genes. The following two equations are available to calculate P_i:

$$p_{io} = p_{1o}\, p_{i,a} = p_{1o}\, a^{-Ng} \approx p_{1o} \cdot 20^{-700g}[-] \tag{1.24}$$

and the exergy contribution of the ith component can be found by combining Equations 1.23 and 1.24:

$$\begin{aligned} Ex &= RTc_i \ln c_i / \left(p_{1o}\, a^{-Ng} c_{oo}\right) = (\mu_1 - \mu_{1o})c_i - c_i \ln p_{i,a} = (\mu_1 - \mu_{1o})c_i - c_i \ln\left(a^{-Ngi}\right) \\ &= 18.7c_i + 700\ (\ln 20)c_i g_i \left[\mathrm{ML^{-1}T^{-2}}\right] \end{aligned} \tag{1.25}$$

The total exergy can be found by summing up the contributions originated from all components. The contribution by inorganic matter can be neglected as the contributions by detritus and even to a higher extent from the biological components are much higher due to an extremely low concentration of these components in the reference system. The contribution by detritus, dead organic matter, is 18.7 kJ/g times the concentration c_i (in g/unit of volume), while the exergy of living organisms consists of

$$Ex_1^{\text{chem}} = 18.7\ \text{kJ/g times the concentration } c_i\ \text{(g/unit of volume)} \tag{1.26}$$

and

$$Ex_i^{\text{bio}} = RT(700 \ln 20)c_i\, g_i = RT\, 2100\, g_i\, c_i \tag{1.27}$$

$R = 8.34$ J/mole and if we presume a molecular weight of an average 10^5 for the enzymes, we obtain the following equation for Ex_i^{bio} at 300 K:

$$Ex_i^{\text{bio}} = 0.0529\, g_i\, c_i \tag{1.28}$$

where the concentration now is expressed in g/unit of volume and the exergy in kJ/unit of volume, as also presumed in Equation 1.26.

For the entire system the exergy, $Ex^{-\text{total}}$ can be found as

$$Ex^{-\text{total}} = 18.7 \sum_{i=1}^{N} c_i + 0.0529 \sum_{i=1}^{N} c_i\, g_i\, [ML^{-1}T^{-2}] \tag{1.29}$$

where g for detritus ($i = 1$) of course is 0. Table 1.1 gives an overview of the exergy of various organisms expressed by weighting factor β that is introduced to be able to cover the exergy for various organisms in the unit detritus equivalent or chemical exergy equivalent.

$$Ex^{-\text{total}} = \sum_{i=1}^{N} \beta_i c_i \,(\text{as detritus equivalent}) \tag{1.30}$$

The β-value is found on the basis of Equations 1.26 and 1.28 by adding the chemical and biological contributions to calculate the eco-exergy. Detritus has in accordance with Equation 1.30 the β-value $= 1.0$.

The biological exergy is found from the knowledge to the entire genome for the 11 organisms and for a number of other organisms by use of a correlation between various complexity measures and the information content of the genome (see Jørgensen et al., 2005). The values of β for various organisms have been discussed in *Ecological Modelling* by several papers, but the latest published values in Jørgensen et al. (in print) are probably coming closest to eventually truth β-values. The previous β-values were generally lower but the relative values between two organisms have not been changed very much by the recently published β-values. As the β-values have been used consequently as relative measures, the previous results obtained by exergy calculations are therefore still valid. Weighting factors defined as the exergy content relative to detritus (see Table 1.1) may be considered quality factors, reflecting how developed the various groups are and to which extent they contribute to the exergy due to their content of information which is reflected in the computation. This is, completely according to Boltzmann (1905), which gave the following relationship for the work, W, that is embodied in the thermodynamic information (see also Section 1.7).

$$W = RT \ln W \quad [\mathrm{ML^2T^{-2}}] \tag{1.31}$$

where W is the number of possible states, among which the information has been selected. W is, as seen for species, the inverse of the probability to obtain the valid amino acid sequence spontaneously.

Table 1.1 β-Values = eco-exergy content relative to the eco-exergy of detritus (Jørgensen et al., 2005)

Early organisms	Plants	Animals
Detritus	1.00	
Viroids	1.0004	
Virus	1.01	
Minimal cell	5.0	
Bacteria	8.5	
Archaea	13.8	
Protists	(Algae) 20	
Yeast	17.8	
	33	Mesozoa, placozoa
	39	Protozoa, amoebe
	43	Phasmida (stick insects)
Fungi, moulds	61	
	76	Nemertina
	91	Cnidaria (corals, sea anemones, jelly fish)
Rhodophyta	92	
	97	Gastroticha
Prolifera, sponges	98	
	109	Brachiopoda
	120	Platyhelminthes (flatworms)
	133	Nematoda (roundworms)
	133	Annelida (leeches)
	143	Gnathostomulida
Mustard weed	143	
	165	Kinorhyncha
Seedless vascula plants	158	
	163	Rotifera (wheel animals)
	164	Entoprocta
Moss	174	
	167	Insecta (beetles, flies, bees, wasps, bugs, ants)
	191	Coleodiea (sea squirt)
	221	Lepidoptera (butterflies)
	232	Crustaceans
	246	Chordata
Rice	275	
Gymosperms (including pinus)	314	
	310	Mollusca, bivalvia, gastropodea
	322	Mosquito
Flowering plants	393	
	499	Fish
	688	Amphibia
	833	Reptilia
	980	Aves (birds)
	2127	Mammalia
	2138	Monkeys
	2145	Anthropoid apes
	2173	*Homo sapiens*

Notice that the β-values are based on eco-exergy expressed in detritus equivalents per gram of biomass, because by multiplication by the biomass concentration per litre you will obtain the eco-exergy in detritus equivalents per litre.The β-values express therefore the specific eco-exergy, which according to Section 1.6 is equal to RTK. R is 8.34 J/mole K or 8.34 kJ 10^{-8}/g presuming a molecular weight of 10^5. The chemical exergy is of course to be added. If we want to use the unit detritus equivalents instead of kJ, a division by 18.7 is needed. It means that the β-values or the specific eco-exergy in detritus equivalents

$$\begin{aligned} 1 + RTK &= 1 + 8.34 \times 300 \times 10^{-8} \ln 20 \times \text{AMS}/18.7 \\ &= 1 + 4.00 \times 10^{-6} \times \text{AMS} \end{aligned} \tag{1.32}$$

where AMS is the number of amino acids in a coded sequence and $\ln 20 = 3.00$.

Kullback's measure of information covers the gain in information, when the distribution is changed from p_{io} to p_i. Note that K is a specific measure (per unit of matter). K multiplied by the total concentration yields the eco-exergy density and K multiplied with the biomass gives the eco-exergy.

Virus has coded about 2500 amino acids. The β-value is therefore only 1.01. The smallest known agents of infectious disease are short strands of RNA. They can cause several plant diseases and are possibly implicated in enigmatic diseases of man and other animals. Viroid cannot encode enzymes. Their replication relies therefore entirely on enzyme systems of the host. Viroid has typically a nucleotide sequence of 360, which means that they would be able to code for 90 amino acids, although they are not translated. The β-value can therefore be calculated to be 1.0004. It can be discussed of course whether viroid should be considered living material.

Eco-exergy calculated by use of the above-shown equations for ecosystems has some clear shortcomings:

1. We have made some although minor approximations in the equations presented above.
2. We do not know the non-nonsense gene and the details of the entire genome for all organisms.
3. We calculate only in principle the exergy embodied in the proteins (enzymes), while there may be other components of importance for the life processes. These components are contributing less to the exergy than the enzymes and the information embodied in the enzymes control the formation of these other components, for instance hormones. It cannot be excluded that these components will contribute to the total exergy of the system.
4. We do not include the exergy of the ecological network, when we calculate the eco-exergy for an ecosystem. If we calculate the exergy of models, the network will always be relatively simple and the contribution coming from the information content of the network is therefore minor, but the information content of real ecological network may be significant due to their high complexity.

5. We will always use a simplification of the ecosystem, for instance by a model or a diagram or similar. This implies that we only calculate the exergy contributions of the components included in the simplified image of the ecosystem. The real ecosystem will inevitably contain more components which are not included in our calculations.

It is therefore proposed that the exergy found by these calculations should always be considered a *relative minimum exergy index* to indicate that there are other contributions to the total exergy of an ecosystem, although they may be of minor importance. In most cases, however, a relative index is sufficient to understand and compare the reactions of ecosystems, because the absolute exergy content is irrelevant for the reactions. It is in most cases the change in eco-exergy which is of importance to understand the ecological reactions.

The weighting factors presented in Table 1.1 have been applied successfully in several structurally dynamic models and furthermore in many illustrations of the maximum exergy principle, that will be presented later in this chapter. The relatively good results in application of the weighting factors, in spite of the uncertainty of their assessment, seem only to be explicable by the robustness of the application of the factors in modelling and other quantifications. The differences between the factors of the microorganims, the vertebrates and invertebrates are so clear, that the uncertainty does only influence the model results minor.

On the other hand, it would be an important progress to get better weighting factors from a theoretical point of view because it would enable us to model the competition between species which are closely related.

There is no doubt that the right estimation of β-values should be based on the number of proteins that are controlling the processes in the cells of various organisms. The knowledge to all human genomes is available today, and it has been found that the number of genes carrying non-repetitive information is not 250,000 as previously applied but rather about 40,000. On the other hand, it is also clear that the number of amino acids controlled by one gene is more than 700 for *Homo sapiens*. The total information in the genes is, under all circumstances, applied in the latest calculation in Table 1.1 of the β-values. The high information content can be explained by the number of amino acids per gene, which for some genes are as high as 38,000 (Hastie, 2001). The β-values in Table 1.1 can be considered the best estimation in the year 2008. The primitive cell is indicated with a β-value of 5.0, while 5.8 was applied previously. Recent results have determined that the oceanic bacterium SAR 11 only has about 1354 genes determining about 948,000 amino acids which would correspond to a β-value of 4.88. It is therefore correct to use 5.0 in Table 1.1.

The importance of the proteins can be seen from the intensive analytical work to find the composition of the human proteins—or as they are called now, the genetic determined proteins, proteoms, a word which is used to underline that the genetic produced proteins are of particular importance (Haugaard Nielsen, 2001). The great interest for the proteoms are due to their control of the life processes. Many proteoms may be the medicine of the future. The application of enzymes in industrial productions is just in its infancy, as there is an enormous potential to control many more industrial processes by enzymes.

The key to find better β-values is the proteoms on the one side. On the other side, our knowledge about the number of proteoms in various organisms is very limited—more limited than for the number of information genes. It may be possible, however, to put together our knowledge about non-nonsense genes, the overall DNA content, the limited knowledge about the number of proteoms and the evolution tree, and see some pattern which could be used to give better but still very approximate β-values at this stage. For *Homo sapiens* it is presumed that 200,000 proteoms are produced by the cells and that they contain about 15,000 amino acids in average (Haugaard Nielsen, 2001). It would give a biological exergy (see Equations 4.21 and 4.22 which is $RT\ (\ln 20)\, 3 \cdot 10^9\ c_i$, and with a molecular weight of 100,000, the β-value would be 12,275, or considerably higher than the value in Table 1.1. The numbers used in these calculations are, however, still very uncertain. The other values in the table may, however, be changed similarly.

1.10 WHY LIVING SYSTEMS HAVE SUCH A HIGH LEVEL OF EXERGY?

What is life? Most scientists would agree that life is

1. the ability to metabolite, that is to draw nutrients from the environment, and convert it into energy, useful biochemical compounds and excrete waste products
2. the ability to reproduce.

These two abilities are rooted in an enormous amount of (useful) information, which is able to control the processes needed to metabolite and reproduce. The β-values are accounted for the information embodied in organisms.

A frog of 20 g will have an exergy content of $20 \times 18.7 \times 688\,\text{kJ} \approx 257\,\text{GJ}$, while a dead frog will have an exergy content of 374 kJ only, although they have the same chemical composition, at least a few seconds after the frog has died. The difference is rooted in the information, or rather the difference in the useful information. The dead frog has the information a few seconds after its death (the amino acid composition has not yet been decomposed), but the difference between a live frog and a dead frog is the ability to utilize the enormous information stored in the genes and the proteoms of the frog. So, the difference in eco-exergy between a live frog and a dead frog may define life as characterized by the ability of utilizing the information in the amino acid sequence.

The amount of information stored in a frog is really surprisingly high. The number of amino acids placed in the right sequence is 84,000,000 and for each of these 84,000,000 amino acids, the number of possibilities is 20. This amount of information is able to ensure reproduction and is transferred from generation to generation which ensures that the evolution can continue because what is already a favourable combination of properties is conserved through the genes.

Because of the very high number of amino acids, 84,000,000, it is not surprising that there will always be a minor difference from frog to frog in the amino acid sequence. It may be a result of mutations or of a minor mistake in the copying process. This variation is important because it gives possibilities to "test" which amino acid sequence gives the best result with respect to survival and growth. This variability has been emphasized by Darwin as a prerequisite for his theory. The best—representing the most favourable combination of properties will offer the highest probability of survival and give most growth and the corresponding genes will therefore prevail. Survival and growth mean more eco-exergy and that a bigger distance to thermodynamic equilibrium is the result. Eco-exergy could therefore be used as a thermodynamic function which could be used to quantify Darwin's theory. This is the idea behind the application of structurally dynamic models.

1.11 TOWARDS A CONSISTENT ECOSYSTEM THEORY

The properties of ecosystems can only be revealed by the use of a pluralistic view. It is therefore not surprising that there are a few different ecosystem theories published in the scientific literature. It is on the other hand necessary to try to unite the theories and examine if they are tied up in contradictions or form a pattern that can be used to give a better understanding of the nature of ecosystems and to solve the global environmental problems. The goal is to give a common framework of reference for *further* development of a more profound and comprehensive ecosystem theory than the one we are able to present today. The pattern should serve as a "conceptual diagram", which can be used as the basis for further discussion of how ecosystems behave.

We are still in an early stage of an ecosystem-theoretical development and it may be argued that this attempt is premature, but the experience from modelling has taught us that it is better to conclude one's thoughts in a conceptual diagram at an early stage and then be ready to make changes than to let all modelling efforts wait until all details are known, as this will never be the case due to the immense complexity of nature (Jørgensen, 2002). Moreover, recent development in ecosystem theory has made it possible to conclude that the theories presented here are indeed consistent and supplementary. This recent development will be presented in this section and Section 1.12.

The centre of the presented pattern below is the tentative fourth law of thermodynamics or ecological law of thermodynamics (ELT), but it cannot be excluded that other formulation of this law could be the core of an ecosystem theory. What can we conclude from this (tentative) law about ecosystem properties? Can we, as it is known from physics, formulate a limited number of laws and explain a very large number of observations (Jørgensen, 2002; Jørgensen and Svirezhev, 2004)? This question has been answered with a clear "yes" in this volume. The recent development in system ecology represents a paradigm shift. The paradigm that is now receding has dominated our culture for several hundred years. It views the universe as a mechanical system composed of elementary building blocks. The new paradigm is based on a holistic world view. The world is seen as an integrated whole and recognizes the fundamental interdependence of all phenomena.

The tentative ecological law of thermodynamics (ELT) gives information on which of the many possible processes will be realized as a result of a competition among more possibilities, than the flow of exergy can accomplish. Ulanowicz (1997) has introduced the expression "the propensity of ecosystems" to stress that ecosystems and their forcing functions encompass many random components and they have an enormous complexity which makes accurate predictions impossible. Propensities are weighted or conditional probabilities that are inherent features of changing situations or occasions rather than absolute properties of relational processes and components. Ever since the quantum mechanics introduced indeterminacy, it has become increasingly easy to recognize that we do actually live in a world of propensities, with an unfolding process of realizing possibilities and creating new possibilities. It seems therefore advantageous to include this expression in the formulation of the pattern of ecosystem theories, and also in the formulation of the tentative fourth law of thermodynamics.

The tentative law asserts therefore, according to the latest formulation, that: A system that receives a through-flow of exergy (high-quality energy) will try to utilize the exergy flow to move away from thermodynamic equilibrium, and if more combinations of components and processes are offered to utilize the exergy flow, the system will select the organization that gives the system as much exergy content (storage) as possible, that is maximizes dEx/dt.

A few very competent ecologists have expressed preference for a formulation where the flow of exergy is replaced by a flow of free energy, which of course is fully acceptable and makes the formulation closer to classic thermodynamics. However, eco-exergy can hardly be replaced by free energy because it is a free energy *difference* between the system and the same system at thermodynamic equilibrium. The reference state is therefore different from ecosystem to ecosystem, which is considered in the definition of eco-exergy. In addition, free energy is not a state function far from thermodynamic equilibrium—just consider the immediate loss of eco-exergy when an organism dies. Before death the organism has high eco-exergy because it can utilize the enormous information that is embodied in the amino acid sequence of the enzymes, which are controlling the life processes. At death the organism looses immediately the ability to use this information, which therefore becomes worthless. The role of information in the evolution will be further discussed in Chapter 2.

The support for the validity of the tentative law in its present formulation is strong and may be summarized in the following three points:

1. It may be considered a translation of Darwin's theory to thermodynamics and is consistent with the basic, thermodynamic laws. The selected organization is the one which offers most "survival" that can be measured as exergy. The selection is in accordance with the latest formulations of Darwin's theory still taking place on the levels of species. The species are surviving, growing and fighting for the resources. All the species are, however, connected in an ecological, cooperative, synergistic network and are dependent on each other. The survival is under the prevailing conditions, which include the presence of all the components in the ecological network. All the species in the ecological network are influencing all the other species. The result is therefore that the entire ecological network gets as much survival and therefore eco-exergy as possible under the prevailing conditions.
2. The application of the hypothetical law in models gives (many) results that are consistent with ecological observations (see Jørgensen, 2002; Jørgensen and Svirezhev, 2004).

3. Many ecological observations can be explained by the presented hypothesis (see Jørgensen et al., 2000; Jørgensen, 2002; Jørgensen and Svirezhev, 2004; Jørgensen et al., 2007).

Below are presented a few case studies from Jørgensen (1997, 2002) and Jørgensen et al. (2000) supporting the presented exergy storage hypothesis, but maximum power or ascendency could also have been applied. More examples can be found in the above-given references.

Size of genomes

In general, biological evolution has been towards organisms with an increasing number of genes and diversity of cell types (Futuyma, 1986). If a direct correspondence between free energy and genome size is assumed, this can reasonably be taken to reflect increasing exergy storage accompanying the increased information content and processing of "higher" organisms.

Le Chatelier's Principle

The exergy-storage hypothesis might be taken as a generalized version of "Le Chatelier's Principle". Biomass synthesis can be expressed as a chemical reaction:

energy + nutrients = molecules with more free energy (exergy) and organization + dissipated energy.

According to Le Chatelier's Principle, if energy is put into a reaction system at equilibrium, the system will shift its equilibrium composition in a way to counteract the change. This means that more molecules with more free energy and organization will be formed. If more pathways are offered, those giving the most relief from the disturbance (displacement from equilibrium) by using the most energy, and forming the most molecules with the most free energy, will be the ones followed in restoring equilibrium.

The sequence of organic matter oxidation

The sequence of biological organic matter oxidation (e.g., Schlesinger, 1997) takes place in the following order: by oxygen, by nitrate, by manganese dioxide, by iron (III), by soleplate, and by carbon dioxide. This means that oxygen, if present, will always outcompete nitrate which will outcompete manganese dioxide, and so on. The amount of exergy stored as a result of an oxidation process is measured by the available kJ/mole of electrons which determines the number of adenosine triphosphate (ATP) molecules formed. ATP represents exergy storage of 42 kJ/mole. Usable energy as exergy in ATPs decreases in the same sequence as indicated above. This is as expected if the exergy-storage hypothesis (ELT) were valid (Table 1.2). If more oxidizing agents are offered to a system, the one giving the highest storage of free energy will be selected.

In Table 1.2, the first (aerobic) reaction will always outcompete the others because it gives the highest yield of stored exergy. The last (anaerobic) reaction produces methane; this is a less complete oxidation than the first because methane has greater exergy content than water.

Table 1.2 Yields of kJ and ATPs per mole of electrons, corresponding to 0.25 moles of CH_2O oxidized (carbohydrates)

Reaction	kJ/mole e^-	ATPs/mole e^-
$CH_2O + O_2 = CO_2 + H_2O$	125	2.98
$CH_2O + 0.8\,NO_3^- + 0.8\,H^+ = CO_2 + 0.4\,N_2 + 1.4\,H_2$	119	2.83
$CH_2O + 2\,MnO_2 + H^+ = CO_2 + 2\,Mn^{2+} + 3\,H_2O$	85	2.02
$CH_2O + 4\,FeOOH + 8\,H^+ = CO_2 + 7\,H_2O + Fe^{2+}$	27	0.64
$CH_2O + 0.5\,SO_4^{2-} + 0.5\,H^+ = CO_2 + 0.5\,HS^- + H_2O$	26	0.62
$CH_2O + 0.5\,CO_2 = CO_2 + 0.5\,CH_4$	23	0.55

Note: The released energy is available to build ATP for various oxidation processes of organic matter at pH = 7.0 and 25°C.

Formation of organic matter in the primeval atmosphere

Numerous experiments have been performed to imitate the formation of organic matter in the primeval atmosphere on Earth 4 billion years ago (Morowitz, 1968). Energy from various sources was sent through a gas mixture of carbon dioxide, ammonia and methane. There are obviously many pathways to utilize the energy sent through simple gas mixtures, but mainly those forming compounds with rather large free energies (amino acids and RNA-like molecules with high exergy storage decomposed when the compounds are oxidized again to carbon dioxide, ammonia and methane) will form an appreciable part of the mixture (according to Morowitz, 1968).

Photosynthesis

There are three biochemical pathways for photosynthesis: (1) the C3 or Calvin–Benson cycle, (2) the C4 pathway, and (3) the crassulacean acid metabolism (CAM) pathway. The latter is least efficient in terms of the amount of plant biomass formed per unit of energy received. Plants using the CAM pathway are, however, able to survive in harsh, arid environments that would be inhospitable to C3 and C4 plants. CAM photosynthesis will generally switch to C3 as soon as sufficient water becomes available (Shugart, 1998). The CAM pathways yield the highest biomass production, reflecting exergy storage under arid conditions, while the other two yield the highest net production (exergy storage) under other conditions. While it is true that a gram of plant biomass produced by the three pathways has different free energies in each case, in a general way improved biomass production by any of the pathways can be taken to be in a direction that is consistent, under the conditions, with the exergy-storage hypothesis (ELT).

Leaf size

Givnish and Vermelj (1976) observed that leaves optimize their size (thus mass) for the conditions. This may be interpreted as meaning that they maximize their free-energy content. The larger the leaves the higher their respiration and evapo-transpiration, but the more solar radiation they can capture. Deciduous forests in moist climates have a leaf-area index (LAI) of about 6 (see also Section 2.4). Such an index can be predicted from the hypothesis of highest possible leaf size, resulting from the trade-off between having leaves of a given size versus maintaining leaves of a given size (Givnish and Vermelj, 1976). Size of leaves in a given environment depends on the solar radiation and humidity regime, andwhile, for example, Sun and shade leaves on the same plant would not have equal exergy contents, and in a general way leaf size and LAI relationships are consistent with the hypothesis of maximum exergy storage.

Biomass packing

The general relationship between animal body weight, W, and population density, D, is $D = A/W$, where A is a constant (Peters, 1983). Highest packing of biomass depends only on the aggregate mass, not on the size of individual organisms. This means that it is biomass rather than population size that is maximized in an ecosystem, as density (number per unit area) is inversely proportional to the weight of the organisms. Of course, the relationship is complex. A given mass of mice would not contain the same exergy or number of individuals as an equivalent weight of elephants. Also, genome differences (Example 1) and other factors would figure in. Later, we will discuss exergy dissipation as an alternative objective function proposed for thermodynamic systems. If this were maximized rather than storage, then biomass packing would follow the relationship $D = A/W^{0.65-0.75}$ (Peters, 1983). As this is not the case, biomass packing and the free energy associated with this lend general support for the exergy-storage hypothesis.

Cycling

If a resource (for instance, a limiting nutrient for plant growth) is abundant, it will typically recycle faster. This is a little strange, because recycling is not needed when a resource is non-limiting. A modelling study (Jørgensen and de Bernardi, 1997) indicates that free-energy storage increases when an abundant resource recycles faster. The result is shown in Figure 1.3. The ratio, R, of nitrogen (N) to phosphorus (P) cycling which gives the highest exergy is plotted in a logarithmic scale versus log(N/P). The plot in Figure 1.3 is also consistent with empirical results (Vollenweider, 1975). Of course, one cannot "inductively test" anything with a model, but the indications and correspondence with the data do tend to support in a general way the exergy-storage hypothesis. The cycling ratio giving the highest ascendency is also correlated similarly to the N/P ratio (personal communication with R. Ulanowicz). In the light of the close relationship between exergy and ascendency, this result is not surprising (see above, Jørgensen, 1995a, Ulanowicz 1997).

Structurally dynamic modelling

Dynamic models whose structure changes over time are based on non-stationary or time-varying differential or difference equations. We will refer to these as *structurally*

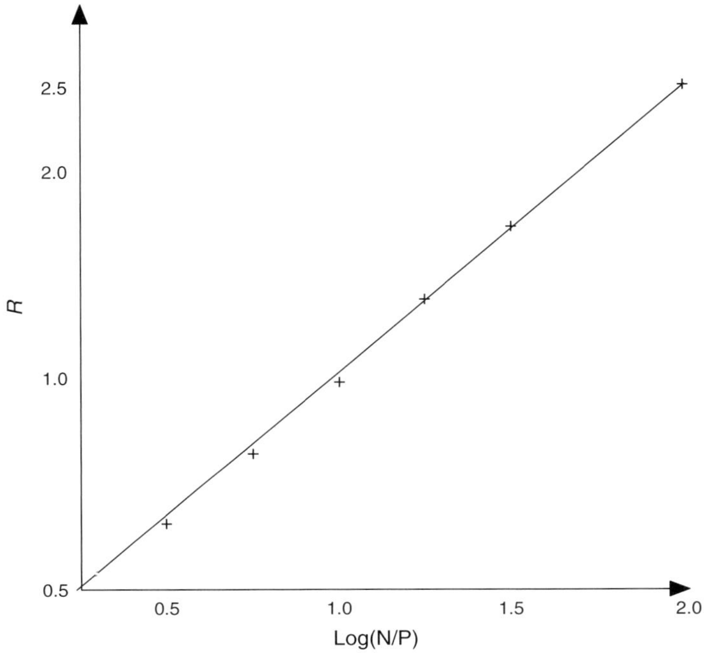

Figure 1.3 Log–log plot of the ratio of nitrogen to phosphorus turnover rates, *R*, at maximum exergy versus the logarithm of the nitrogen/phosphorus ratio, log *N*/*P*. The plot is consistent with Vollenweider (1975).

dynamic models. A number of such models, mainly of aquatic systems (Jørgensen, 1986, 1988, 1990, 1992a,b; Jørgensen and Padisak, 1996; Coffaro et al., 1997; Jørgensen and de Bernardi, 1997, 1998), but also as population dynamic models (Jørgensen, 2002) and terrestrial systems (Jørgensen and Fath, 2004), have been investigated to see how structural changes are reflected in free-energy changes. The technicalities of parameter fitting aside, this overall result means that the system structure must change if its eco-exergy storage is to be continually maximized. Changes in parameters, and thus system structure, not only reflect changes in external boundary conditions, but also mean that such changes are necessary for the ongoing maximization of exergy. For all models investigated along these lines, the changes obtained were in accordance with actual observations (see references). These studies therefore affirm, in a general way, that systems adapt structurally to maximize their content of exergy. The shortcomings of assessing the exergy content of an ecosystem have been discussed above. At least in models the applicability of the exergy calculations has shown their more practical use, which can be explained by a robustness in the model and eco-exergy calculations. It is most probably important to have different weighting factors for organisms that are developed radically different and it is probably also important that the model focuses on the very problem that causes the structural changes, but whether the exergy contributions calculated are too high or too low is not important for a description of the structural changes.

It is noteworthy that Coffaro et al. (1997), in his structural dynamic model of the Lagoon of Venice, did not calibrate the model describing the spatial pattern of various macrophytes species such as *Ulva* and *Zostera*, but used exergy-index optimization to estimate parameters determining the spatial distribution of these species. He found good accordance between observations and model, as was able by this method *without* calibration to explain more than 90% of the observed spatial distribution of various species of *Zostera* and *Ulva*.

We need a number of different complementary approaches to explain ecosystems which are not surprising as much simpler physical phenomena; light, for instance, needs two different descriptions, namely as waves and as particles. Several ecosystem theories have been presented in the scientific literature during the last 2–3 decades. At the first glance they look very different and seem to be inconsistent, but a further examination reveals that they are not so different and that it should be possible to unite them in a consistent pattern. This was already stated in the book *Integration of Ecosystem Theories: A Pattern*, in its first edition in 1992. Now it has been published as a third edition (Jørgensen, 2002). This unification of the various ecosystem theories has been widely accepted among system ecologists since 1998/1999, but as a result of two meetings in year 2000—one in Italy, Porto Venere late May, and one in Copenhagen, early June in conjunction with an American Society of Limnology and Oceanography (ASLO) meeting—it can now be concluded that a consistent pattern of ecosystem theories has been formed. Several system ecologists agreed on the pattern presented below, as a working basis for further development in system ecology. Further steps towards a more comprehensive and more consistent ecosystem theory are taken with this volume. This is of utmost importance for the progress in system ecology, because with a theory in hand it will be possible to explain many rules that are published in ecology and applied ecology which again explain many ecological observations. Moreover, by a good theory in hand it is possible to calculate and predict what otherwise would require a lot of expensive observations and trial and error experiments. We should, in other words, be able to attain the same theoretical basis that characterizes physics: a few basic laws, which can be used to deduce rules that explain observations.

An important contribution to a clear pattern of the various ecosystem theories came from the network approach used often by Patten (see, for instance). Fath and Patten (2001) and Fath et al. (2004) have shown by a mathematical analysis of networks in steady state (representing, for instance, an average annual situation in an ecosystem with close to balanced inputs and outputs for all components in the network) that the sum of through-flows in a network (which is power) is determined by the input and the cycling within the network. The input (the solar radiation) again is determined by the structure of the system (the stored exergy, the biomass). Furthermore, the more structure the more maintenance is needed. Therefore, more exergy must be dissipated, the greater the inputs are. Cycling, on the other hand, means that the same energy (exergy) is utilized better in the system, and therefore more biomass (exergy) can be formed without an increase of the inputs. It has been shown previously that more cycling means increased ratio of indirect to direct effects, while increased input doesn't change the ratio of indirect to direct effects, as both effects increase by the same factor.

Fath and Patten (2001) used these results to determine the development of various variables used as goal functions (exergy, power, entropy, etc.). An ecosystem is of course not setting goals, but a goal function, or orientor maybe a better word to use in this context, can be used to describe the direction of development an ecosystem will take. Their results can be summarized as follows:

1. Increased inputs (more solar radiation is captured) means more biomass, more exergy stored, more exergy degraded, therefore also higher entropy dissipation, more through-flow (power), increased ascendency, but no change in the ratio of indirect to direct effects or in the retention time for the energy in the system = total exergy/ input exergy per unit of time.
2. Increased cycling implies more biomass, more exergy stored, more through-flow, increased ascendency, increased ratio of indirect to direct effects, increased retention, but no change in exergy degradation.

Almost simultaneously Jørgensen et al. (2000) published a paper which claims that ecosystems show three growth forms:

I. Growth of physical structure (biomass) which is able to capture more of the incoming energy in the form of solar radiation, but also requires more energy for maintenance (respiration and evaporation).
II. Growth of network, which means more cycling of energy and matter.
III. Growth of information (more developed plants and animals with more genes), from r-strategists to K-strategists, which waste less energy but usually carry more information.

These three growth forms may be considered an integration of E.P. Odum's attributes which describe changes in ecosystem associated with development from the early stage to the mature stage. Nine of the most applied attributes associated to the three growth forms should be mentioned:

1. Ecosystem biomass (biological structure) increases.
2. More feedback loops (including recycling of energy and matter) are built.
3. Respiration increases.
4. Respiration relative to biomass decreases.
5. Bigger animals and plants (trees) become more dominant.
6. The specific entropy production (relative to biomass) decreases.
7. The total entropy production will first increase and then stabilize on approximately the same level.

8. The amount of information increases (more species, species with more genes, the biochemistry becomes more diverse).
9. r-strategists are replaced by K-strategists.

Growth form I covers attributes 1, 3 and 7. The biomass increases according to attribute 1, which implies that also the respiration increases, because it costs more exergy to maintain more biomass. This means that also the entropy production will increase.

Growth form II covers 2 and 6. When the network increases, there will be more feedback mechanisms available for regulation of the network. The energy and mass will thereby circle to a higher extent which means that more biomass can be supported with the same total input and output of the eco-exergy.

Growth form III covers the attributes 4, 5, 7, 8 and 9. Bigger and more developed species will take over according to growth form III. It implies more biomass relative to the respiration and while the total entropy production is not changed, the specific entropy production is decreased.

Five of the presented hypotheses to describe ecosystem growth and development are examined with respect to the three growth forms:

A. The entropy production tends to be minimum (this is proposed by Prigogine, 1947, 1955 and 1980, for linear systems at steady non-equilibrium state, not for far from equilibrium systems). It is applied by Mauersberger (1983, 1995) to derive expressions for bioprocesses at a stable stationary state.
B. Natural selection tends to make the energy flux through the system a maximum, so far as compatible with the constraints to which the system is subject (H.T. Odum, 1983, 1988, 1996). This is also called the maximum power principle.
C. Ecosystem will organize themselves to maximize the degradation of exergy (Kay, 1984, 1991).
D. A system that receives a through-flow of exergy will have a propensity to move away from thermodynamic equilibrium, and if more combinations of components and processes are offered to utilize the exergy flow, the system has the propensity to select the organization that gives the system as much stored exergy as possible (Jørgensen and Mejer, 1977, 1979; Jørgensen, 1986, 1988, 1990, 1992a, 2002).
E. Ecosystem will have a propensity to develop towards a maximization of the ascendency (Ulanowicz, 1986).

The usual description of ecosystem development illustrated, for instance, by the recovery of Yellowstone Park after fire, an island born after a volcanic eruption, reclaimed land and so on, is well covered by E.P. Odum (1969a, 1969b, 1971): at first, the biomass increases rapidly which implies that the percentage of captured incoming solar radiation increases and also the energy needed for the maintenance. Growth form I is dominant in this first phase, where not only the exergy stored increases (more biomass, more physical structure to capture more solar radiation), but also the through-flow (of useful energy), exergy dissipation and the entropy production increases due to increased need of energy for maintenance.

Growth forms II and III become dominant later, although an overlap of the three growth forms takes place. When the percentage of solar radiation captured reaches about 80%, it is not possible to increase the amount of captured solar radiation further (due in principle to the second law of thermodynamics).

Further growth of the physical structure (biomass) therefore does not improve the exergy balance of the ecosystem. In addition, all or almost all the essential elements are in the form of dead or living organic matter and not as inorganic compounds ready to be used for growth. The growth form I will therefore not proceed, but growth forms II and III can still operate. The ecosystem can still improve the ecological network and can still replace r-strategists by K-strategists, small animals and plants by bigger ones (Cope's Law: the later descendent may be increasingly larger than their ancestors—for instance, the horse today is much bigger than the horse fossils from 20 to 30 million years ago) and less developed by more developed.

Growth forms II and III require, however, not more exergy for maintenance. Exergy degradation is therefore not increasing but is maintained on a constant level, or expressed differently: specific exergy degradation and specific entropy production are decreasing with growth forms II and III. It is in accordance with Aoki (1998, 2006). He has shown that respiration per biomass or entropy production per biomass is increasing with increasing trophic diversity up to a trophic diversity of 3–4 and afterwards decreasing rapidly with further increase of the trophic diversity up to 7 for aquatic ecosystems. The results by Aoki are in accordance with the actual observations, which have been applied to develop ECOPATH models (Christensen and Pauly, 1993; Christensen, 1995).

The accordance with the five descriptors + specific entropy production and the three growth forms based on this description of ecosystem development is shown in Table 1.3.

Based upon the results, it is possible to formulate the following hypothesis which unites the five hypotheses.

Table 1.3 Accordance between growth forms and the proposed descriptors

	Hypothesis		
	Growth form I	Growth form II	Growth form III
Exergy storage	Up	Up	Up
Power/through-flow	Up	Up	Up
Ascendency	Up	Up	Up
Exergy dissipation	Up	Equal	Equal
Retention time	Equal	Up	Up
Entropy production	Up	Equal	Equal
Exergy/biomass = specific exergy	Equal	Up	Up
Entropy/biomass = specific entropy production	Equal	Down	Down
Ratio indirect/direct effects	Equal	Up	Up

Ecosystem development in all phases will move away from thermodynamic equilibrium and select the components and the organization that yields the highest flux of useful energy through the system and the most exergy stored in the system. This corresponds also to the highest ascendency.

Ecosystem development is accomplished by the three growth forms:

1. Growth of biomass (physical structure) which implies that more exergy is degraded due to an increased demand for maintenance energy. An ecosystem tends according to growth form I to reach the highest possible rate of exergy captured (which is in the order of 80% of the incoming solar radiation) and thereby also of exergy degradation. This growth form is therefore best measured by a determination of the exergy degradation rate.
2. Growth of the number of network linkages and thereby of recycling of matter and energy which implies a better utilization of the incoming energy, and therefore an increase in through-flow and exergy storage without an increase in exergy dissipation. It means that specific exergy degradation and specific entropy production are decreasing.
3. Growth of the number of components in the network and replacement of r-strategists and small organisms with K-strategists and bigger organisms. It implies the same changes take place as observed by growth form II, namely increased through-flow, increased exergy storage and decreased specific exergy degradation and entropy production.

The possibilities for an ecosystem to use growth form number one are limited, because of the amount of elements that is present.

The element present in the smallest amount relative to the demand will stop growth of biomass.

Growth form I could be stopped even before, when it captures about the 80% of the solar radiation that is physically possible. As it will be discussed in Chapter 2, we are very far from the limits for growth forms II and III. The evolution has therefore to a high extent played on these two growth forms: growth form III (more and more complex organisms) and growth form II (more and more different organisms are linked in a more and more complex network).

In the paper by Jørgensen et al. (2000), Figure 1.4 was presented to illustrate the concomitant development of ecosystems, exergy captured (most of that being degraded) and exergy stored (biomass, structure, information). The points in the figures correspond to ecosystems on different stages of development (see Table 1.4).

Debeljak (2001) has shown that he gets the same shape of the curve when he determines exergy captured and exergy stored in managed forest and virgin forest on different stages of development (see Figure 1.5). The exergy captured was determined as in Table 1.4 by measurement of the temperature of the infrared radiation, while the exergy storage was determined by a randomized measurement of the size of all trees and plants. The stages are indicated on the figure, where also pasture is included for

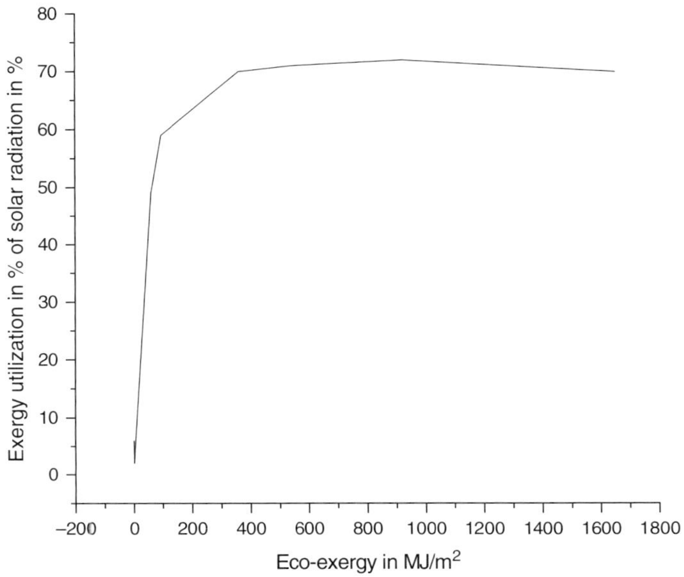

Figure 1.4 The exergy captured (taken from Kay and Schneider, 1992, % of solar radiation) is plotted versus the exergy stored, unit MJ/m^2, calculated from characteristic compositions of the eight focal ecosystems. The values from Table 1.4 are applied to construct this plot. Notice that exergy utilization is parallel (proportional) to the energy absorbed.

Table 1.4 Exergy utilization and storage in a comparative set of ecosystems

Ecosystem	Exergy utilization (%)	Exergy storage (MJ/m^2)
Quarry	6	0
Desert	2	0.75
Clear-cut forest	49	60
Grassland	59	94
Fir plantation	70	360
Natural forest	71	540
Old-growth deciduous forest	72	920
Tropical rain forest	70	1650

comparison. Catastrophic events such as storm or fire may cause destructive regeneration, as described in Holling's cycle; see Jørgensen et al. (2007) and Chapter 7 for a detailed description. Destructive regeneration has happened several times during the evolution.

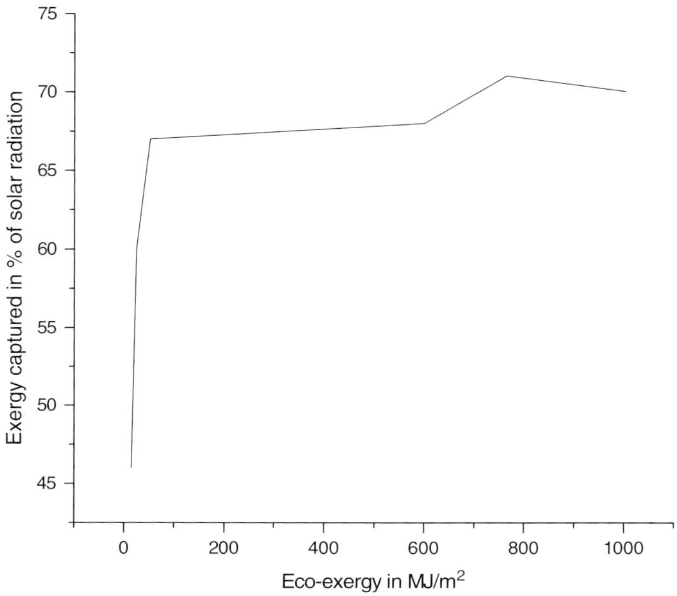

Figure 1.5 The plot shows the result by Debeljak (2001). He examined managed forest and virgin forest in different stages. Gap has no trees, while the virgin forest changes from optimum to mixed to regeneration and back to optimum, although the virgin forest can be destroyed by catastrophic events such as fire or storms. The juvenile stage is a development between the gap and the optimum. Pasture is included for comparison.

1.12 EXERGY BALANCES FOR THE UTILIZATION OF SOLAR RADIATION

In Jørgensen and Svirezhev (2004) the following expression has been shown for eco-exergy gained, *Ex*, as a result of the energy of the incoming solar radiation, *E*-in:

$$Ex = (E\text{-in} - R)\ [K - \ln\ ((E\text{-in} - R)/E\text{-in})] + \mathrm{R} \tag{1.33}$$

where *R* is the difference between the total incoming and outgoing radiation and *K* is Kullback's measure of information. If we introduce the radiation efficiency eff-rad = R/E-in and the exergy efficiency, eff-Ex = Ex/E-in, Equation 1.33 can be reformulated as

$$\text{Eff-Ex} = (1-\text{eff-rad})K + (1 - \text{eff-rad}) \ln(1 - \text{eff-rad}) + \text{eff-rad} \tag{1.34}$$

Eff-Ex is therefore a function of two independent variables, eff-rad and *K*, but is independent on any parameter.

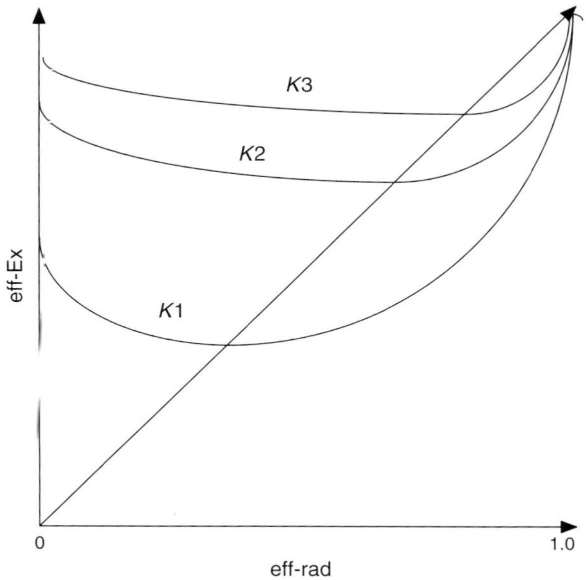

Figure 1.6 eff-ex is plotted versus eff-rad for three information levels, $K3 > K2 > K1$. Increasing Kullback's measure of information implies that the ecosystem will work as an information machine up to a higher eff-rad.

Figure 1.6 shows the relationship between eff-Ex and eff-rad for three different values of *K*. The active surface of an ecosystem as seen in the figure will operate as a classic thermodynamic machine performing mechanical and chemical work, while the active surface will operate as an information machine producing information when *K* is high and eff-rad is not too high. It can easily be seen that when $K = 0$ the system will work as a classic thermodynamic machine, while increasing *K* will imply that the ecosystem will work as an information machine up to a higher and higher eff-rad. When an ecosystem has attained a certain level of eco-exergy/information it will continue to work as an information machine increasing the eco-exergy without an increase of the physical structure.

Seasonal changes of Ex, *R*, eff-Ex, eff-rad and *K* for a forest and for a crop field have been found by Svirezhev et al. (2003). From these results, it can be concluded that

1. the vegetation works as an information machine (Ex $> r$) and eff-Ex $>$ eff-rad in the course of (almost) the entire year.
2. exergy, *R*, eff-Ex, Ex-rad and *K* achieve maxima when the productivity of vegetation is also maximal; it means in the Northern Hemisphere in June–July.

2

Thermodynamics and the evolutionary processes

2.1 EVOLUTION AND CONSTRAINTS

The evolution is often described as a stepwise development from inorganic to simple organic molecules and further on via more complex molecules and simple organisms towards more and more complex organisms that have more and more sophisticated properties. Even the simplest organisms would never have been formed spontaneously, because the probability that the right assemble of the complex biochemical compounds that determine the life processes is formed spontaneously would be so low that in practice it would never happen. The evolution has been possible only because the progressive steps were made one by one—from simple organic molecules to more and more complex molecules that were brought together to form something which after numerous trials and errors could be reproduced by, maybe in the first hand, a very simple method. Later the reproduction method became more complex and more effective, of course, as all the life-determining processes.

Mayr (2001) has pointed out that almost all evolutionary phenomena can be assigned to one or the other of two major evolutionary processes, namely the acquisition and maintenance of adaptedness and the origin and role of organic diversity. Mayr (2001) means that most treatments of evolution are written in a reductionist manner in which all evolutionary phenomena are reduced to the level of the gene. This approach inevitably fails because evolution deals with phenotypes of individuals, with populations and with species.

The evolution would, however, not have been possible without the genes, that is, the ability to transfer information from generation to generation and thereby build on shoulders of the ancestors. The genes, via the complementary copies of mRNA molecules, determine the primary structure of the amino acid sequence of protein molecules—no more any less. The primary structure of proteins determines all of the higher order structures and thus their functions. Some proteins are used as building blocks, others play the role of enzymes and another class of proteins participates in the so-called cell signalling, that is, perceiving impulses (mostly chemical) from outside and transmitting them within the cell. In addition, there are also so-called motor proteins that transform chemical energy into mechanical energy. All these categories of proteins are indispensable for the development of the organism and its survival (Haugaard Nielsen, 2001).

The constraints in the beginning of the evolution were that whenever a primitive but relatively well-functioning assemble of organic molecules was formed, the composition that made the "organism" successful was forgotten after the death of the "organism". The next "organism" would have to start from scratch again. If at least the major part of the well-functioning composition could be remembered, the organisms would be able to improve generation by generation their composition and processes. It would save a lot of eco-exergy not to be forced to start from scratch, but to start immediately from an already workable solution. Constraints imply both problems and challenges and therefore new possibilities—a dualistic interpretation. In the first hand, the constraints are problems, but problems can also be considered to be challenges in the sense that, if the problems are solved, improvements and new possibilities are created. For organisms the eternal problem is to survive. When new life conditions are emerging, the accompanying problem can be solved by changing the properties of the organisms and/or their interactions in an ecological network to become better fitted to the new life conditions. The survival based on the two growth forms, growth of biomass and structure and growth of the ecological networks, is ensured by a steady adaptation to the new prevailing life conditions. But a good solution to the survival under new emergent conditions requires a system to transfer information to the next generation, whereby the knowledge about a good solution is maintained. The constraints (problems) that the information about good solutions has to be transferred to the next generation are solved by the genes that again put new constraints on survival: it is possible to ensure survival only in the light of the competition by use of genes. But the genes have also created new possibilities, because mutations and, later in the evolution, sexual re-combinations and other mechanisms, which will be discussed later in this movement of the book, create a wide range of new possible solutions. Therefore, as shown in Figure 2.1, what starts with constraints and new and better properties of the organisms and/or their ecological network end up as new possibilities through a well-functioning coding system.

The growth of information is as seen different from the growth of biomass and network, first because information transfer is needed to remember the good solutions to the growth of the biomass and the ecological network, second because any information transfer system may be considered as constraints, which, however, as other constraints open up new possibilities.

The evolution can be considered as (almost) infinite shifts between problems (constraints) followed by solutions that create new possibilities but also new problems (constraints) that call for new solutions that create new possibilities and so on—a staircase towards a more and more complex world. Notice, however, that the selection following emergent constraints or problems takes place among phenotypes, while the new and better possibilities are created in the genes.

The development has similarities with communication after the emergence of the language. When the human language was created probably a couple of million years ago in its most primitive form, the language at the first hand was new constraints for humans. They had to learn the language and use it, but once they have mastered the language it also gave new opportunities, because it made it possible to discuss co-operation and a detailed better hunting strategy for example, which would increase the possibility for

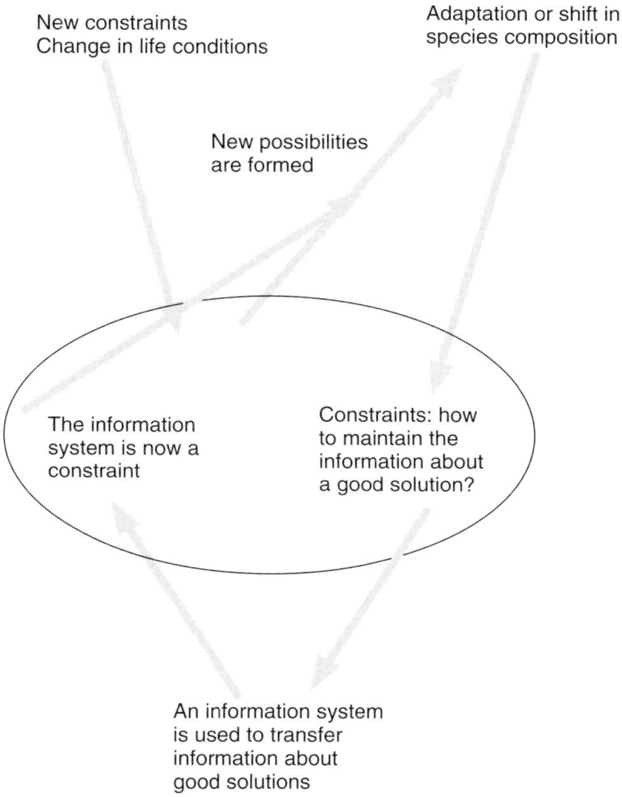

Figure 2.1 The life conditions are currently changed and have a high variability in time and space. This creates new challenges (problems) to survival. The organisms can adapt or a shift to other better fitted species can take place. This requires an information system that is able to transfer the information about good solutions to the coming generations of organisms. Consequently, an information system is very beneficial, but it may also be considered as a new constraint that however opens up for new possibilities.

survival. But it created new constraints: how to communicate more independently of time and space? The written language was developed to solve this problem of transfer of messages. But to learn to read and write were new constraints to humans that, however, also opened up many new possibilities of expressing new ideas and thoughts and thereby made it possible to move further away from thermodynamic equilibrium, which called for a wider and faster communication that was solved by internet and so on.

It has been widely discussed whether the evolution has been gradual or by jumps. The assumption today (see Jensen, 2005) is that both descriptions are valid. During a period with stable conditions, a gradual evolution is dominant. When sudden changes in the conditions for life, for instance sudden climatic conditions, occur, evolution by jumps may be more dominant.

2.2 EVOLUTION AND THE GENES

We will use an example to illustrate the enormous evolutionary power of the genes to transfer information from generation to generation. If a chimpanzee would try to write this book by randomly using a computer keyboard, the chimpanzee would not have been able to write the volume even if it had started at the big-bang 15 billion years ago: but if we could control the first attempt to write the volume and use the signs that were correct for the second round and so on, then 1/40 of the volume would be correct in the first round (assuming 40 different signs), $(39 \times 39)/(40 \times 40)$ would still be incorrect after the second round, $(39 \times 39 \times 39)/(40 \times 40 \times 40)$ after the third round and so on. After 500 rounds—which may take a few years—there would be only 5 "printed" errors left, if we presume that 1 volume contains 600,000 signs. To write the volume would probably require 600,000 seconds or approximately about 1 week. To make 500 rounds would then take 500 weeks or about 9 years.

The biochemistry of organisms is determined by the composition of a series of enzymes that again are determined by the genes. Successful organisms will be able to get more offsprings than less successful organisms and as the gene composition is inherited, the successful properties will be more and more represented generation after generation. This explains that the evolution has been towards more and more complex organisms, which have new and emerging properties; see for instance Figure 2.2, where an evolution index is plotted versus time. The index is found as the product of the number of marine families and the β-value of the most developed organism at a given time. It is presumed generally that the development of the number of species follows the same pattern as the development of the number of families. The β-value has been

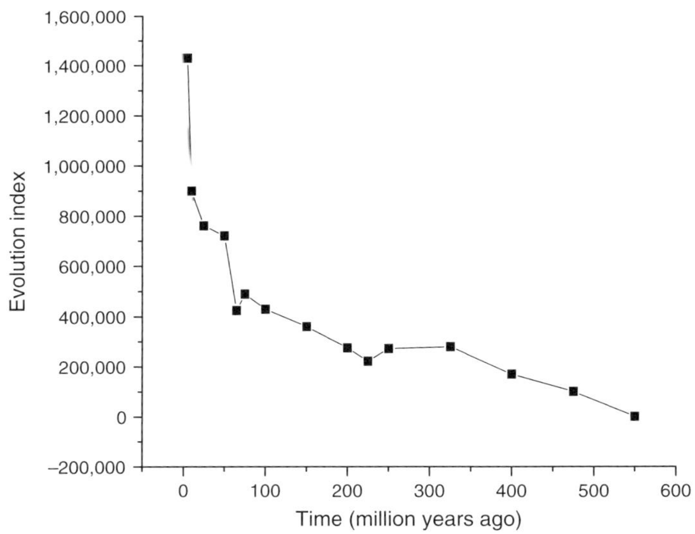

Figure 2.2 An evolutionary index is plotted versus time. The calculation of the index is explained in the text.

Table 2.1 Time of emergence applied to calculate the evolution index shown in Figure 2.2

Ma	Animals emergent
500–550	A wide spectrum of invertebrates
475	Primitive fish
400	Fish
375	Amphibians
300	Reptiles
200	Mammals
30	Monkeys
10	Apes
1	Human

presented in Jørgensen et al. (2005). The applied β-values are shown in Table 1.1, Chapter 1. Table 2.1 gives the time of emergence applied in Figure 2.2.

The genetic code is a language or an alphabet. In the first hand, it is a constraint on the living organisms that has to follow the biochemical code embodied in the genes. As an alphabet is a constraint for an author (he has to learn it and he is forced to use it if he wants to express his thoughts), so the genetic code is a constraint for the living organisms. But as the alphabet gives a writer almost unlimited opportunities to express thoughts and feelings, so the genetic code has given the living organisms opportunity to evolve, becoming more and more complex, more and more creative and more and more adaptive to constraints that are varying in time and space. The need for heritage of useful properties has, in the first hand, been constraints. The genetic code has, however, not only solved the problem associated with these constraints, but it has also been able to give the living organisms new emergent properties and enhanced the evolution.

The evolution is formed by the constraints as the challenges for the organisms are originated in steadily changed life conditions. The organisms have been forced to provide most possible growth by a wide spectrum of changeable life conditions. Due to mutations and, later in the evolution, sexual recombinations, new solutions to the survival in a changeable world have been provided and if there, among the new solutions, were better solutions than the previous ones, the information has been stored in the genes and can be used in the future. This image of the evolution is in accordance with Monod's description of evolution as a result of chance and necessity. The chance or random element of the evolution is the steadily varying life conditions and the necessity is the survival, because without survival there would be no continuation.

The genetic code contains the combination of four amino bases in blocks of four. The genetic code opens therefore for $4 \times 4 \times 4 \times 4$ combinations, but the code is used to select slightly over 20 amino acids, which implies that the code contains redundant signals.

It does not matter anything for the efficiency of the code system; but it seems to indicate that the genetic code itself has an element of randomness.

New constraints are needed to give the evolution a kick from time to time. It has, therefore, probably been an advantage for the evolution that the Earth has been witness

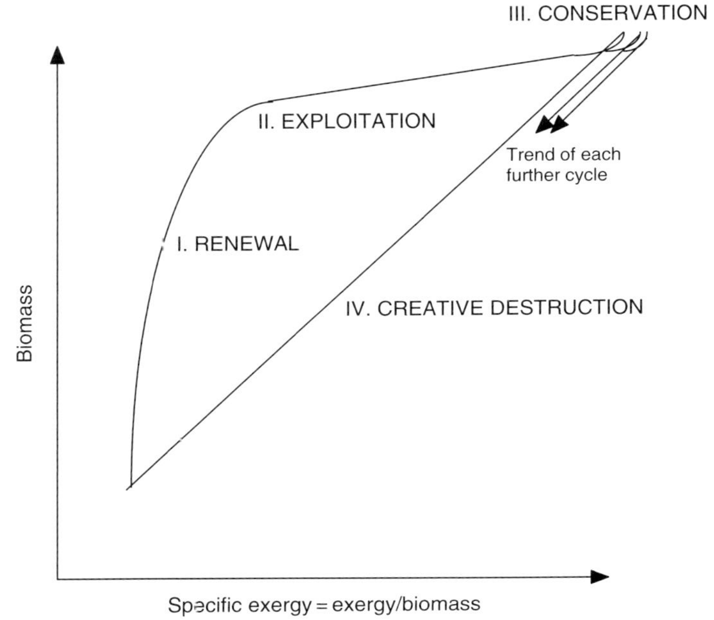

Figure 2.3 Holling's four phases of ecosystems, described in terms of biomass vs. specific exergy. The presentation is inspired by Ulanowicz (1997).

to several enormous nature catastrophes, as for instance, when an asteroid probably hit the Earth 65 million years ago and thereby created new conditions and therefore new challenges. New solutions may often not have a chance as long as old solutions are dominant. Only elimination of old solutions can give a new and different start. The evolution is dependent on catastrophes from time to time—volcanic eruptions, hurricanes and sudden climatic changes. Compare with Holling's cycle (Holling, 1986); see Figure 2.3.

About 5 million years ago the climate in Africa changed dramatically. The precipitation was reduced significantly and the rain forest in East Africa was replaced mainly by savanna. This was a new challenge to the apes: it would be more beneficial for the apes in the savanna to be bipedal. That started the evolution towards the modern man, *Homo sapiens*. The evolution towards a bigger brain volume was probably also or at least partially started randomly: man became carnivorous and the brain growth needed a wide spectrum of amino acids and at the same time hunting required a bigger brain because successful hunting required a co-ordinated team work—the tribe formed therefore an information network to facilitate the communication (exchange of information) among the members of the tribe and the hunting team.

The relationship between the random changes in the life conditions or the constraints and the evolution means that if we would repeat the evolutionary process, for instance, since the Cambrian Explosion, it would inevitably follow the same ecological principles—the same ecosystem theoretical propositions, but it is not at all certain that the final results would be the same due to the role of randomness. This issue is touched on further by the discussion of Drake's equation in the Second Movement.

2.3 THE ARROW OF THE EVOLUTION

The entire evolution from the big-bang via formation of the Sun and the Earth to the cultural-technological level of *Homo sapiens* of today has of course followed the second law of thermodynamics. It means that the evolution—not surprisingly—is irreversible. The irreversibility principle according to the second law of thermodynamics is a necessity for the evolution, but not sufficient. The evolution has one direction: towards a higher and higher level of complexity, order and information. If the processes that make up the evolution would be reversible, the evolution would not be directional. The cosmic evolution that is discussed in the Second Movement from super-concentrated matter to particles to atoms to galaxies to stars and planets illustrates how a hierarchy has been formed and enriched as a consequence of the directionality.

The biological evolution from simple organic molecules to complex organic molecules to prokaryotic cells to eukaryotic cells to invertebrates to vertebrates and finally to *Homo sapiens* via mammals, monkeys and apes shows the same direction of the evolutionary arrow: higher and higher level of complexity, order and information.

Biological systems differ from physical systems by the increased complexity inherent in their development (Tiezzi, 2006).

Order and information can be measured by eco-exergy (see Chapter 1). It is therefore hypothesized that eco-exergy of the Earth will show an increasing function of time. It is attempted to make some additional calculations of the important steps in the cosmic evolution to illustrate this increase, for instance from interstellar clouds to stars; but the emphasis will be on the biological steps from simple organic molecules to *Homo sapiens*. The mechanism proposed for the evolution is that the eco-exergy is steadily increasing with some temporary decline of the eco-exergy levels when major catastrophes (completely new constraints) occur. The eco-exergy strives towards a higher and higher level; but the forcing functions may change the conditions and therefore the eco-exergy level, too. Therefore, it is most correct to use the phrase that "the development uses the pathways giving the highest eco-exergy under the *prevailing conditions*". To obtain "the highest eco-exergy" is a necessity in the Monod's sense while the *prevailing conditions* represent the chance—the random element.

The search for a natural explanation for the origin of ordered structure has challenged the humans since the ancient time. From our thermodynamic considerations in Chapter 1, it is clear that the origin is an energy source from outside the system that has to be open. An energy inflow is both a necessary condition for ordered structure because of the second law of thermodynamics (it will always cost energy to maintain a structure) and a sufficient condition, because an inflow of energy will always form an ordered (dissipative) structure. It was shown by Morowitz (1968) that an inflow of energy will make at least one cycle of matter, that is an ordered structure.

It is interesting that while the stored eco-exergy is increasing as a result of the evolutionary processes—more structure, more order, more information—also the eco-exergy flow density (per unit of mass), which will be calculated throughout the Second Movement for the various steps of the evolution, is increasing. It implies that the system obtains more dynamics and is able to move faster away from thermodynamic equilibrium. It is in close agreement with the consistency between the maximum power and the maximum eco-exergy hypotheses (see Chapter 1).

The values of the most important physical constants (Barrow, 2003) have recently got much attention. Let us mention the most crucial of these physical constants, for instance b defined as the ratio between the mass of the electron and the proton, 1/1836, and the fine structure constant, $\alpha = 2\pi hc/e^2 = 1/137$, where h is Planck's constant, c is the velocity of light and e is the charge of the electron. If b is increased too much, there could be no ordered molecular structure and if b exceeds $0.005\,\alpha^2$, there would have been no stars. The biologically vital elements like carbon and nitrogen cannot exist unless the strong nuclear force is $> 0.3\,\alpha^{1/2}$. The energy of the possible states of carbon-12 is crucial for the formation of life. If the energy of the carbon-12 states were a little different from the actual values, life at least in the form known on the Earth would have been impossible. Why the physical constants have values (and the possible ranges are reasonably narrow) that make it possible to develop life in the universe? I don't believe it will ever be possible to answer this question, but it may be possible to argue that the universe is created to bear life. Is life the real or only meaning behind the creation of the universe? No answer. Of course, it will inevitably lead to religious questions to continue this discussion.

2.4 THE TEMPERATURE RANGE NEEDED FOR CARBON-BASED LIFE PROCESSES

The input of energy for ecosystems is in the form of the solar photon flux. This comes as small portions (quanta) of energy ($= h\nu$, where h is Planck's constant and ν is the frequency) which imply that the exergy at first can only be utilized at molecular (lowest) levels in the biological hierarchy. The appropriate atoms or molecules must be transported to the place where order is created. Diffusion processes through a solid are extremely slow, even at room temperature. The diffusion of molecules through a liquid is about three orders of magnitude faster than in a solid at the same temperature. Diffusion coefficients for gases are ordinarily four orders of magnitude greater than for liquids. This implies that the creation of order (and also the inverse process, disordering) is much more rapid in liquid and gaseous phases than in solids. The temperature required for a sufficiently rapid creation of order is consequently considerably above the lower limit mentioned above, 2.726 K. As far as diffusion processes in solids, liquids and gases are concerned, gaseous diffusion allows the most rapid mass transport. However, many molecules on Earth that are necessary for ordinary carbon-based life do not occur in a gaseous phase, and liquid diffusion, even though it occurs at a much slower rate, is of particular importance for biological ordering processes.

The diffusion coefficient increases significantly with temperature. For gases, the diffusion coefficient varies with temperature approximately as $T^{3/2}$ (Hirschfelder et al., 1954), where T is the absolute temperature. Thus, we should look for systems with the high-order characteristic of life, at temperatures considerably higher than 2.726 K; see also Section 1.5. The reaction rates for biochemical-anabolic processes on the molecular level are highly temperature-dependent (see Straskraba et al., 1997). The influence of temperature may be reduced by the presence of reaction-specific enzymes, which are proteins formed by anabolic processes. The relationship between the absolute temperature, T, and the reaction rate coefficient, k, for a number of biochemical processes can be expressed by the following general equation (see any textbook in physical chemistry):

$$\ln k = b - A/R \times T \tag{2.1}$$

where A is the so-called activation energy, b is a constant and R is the gas constant. Enzymes are able to reduce the activation energy (the energy that the molecules require to perform the biochemical reaction). Similar dependence of the temperature is known for a wide spectrum of biological processes, for instance growth and respiration. Biochemical and biological kinetics point, therefore, towards ecosystem temperatures considerably higher than 2.726 K.

The high efficiency in the use of low-entropy energy at the present "room temperature" on Earth works hand in hand with the chemical stability of the chemical species characteristic of life on Earth. Macromolecules are subject to thermal denaturation. Among the macromolecules proteins are most sensitive to thermal effects. The constant breakdown of proteins leads to a substantial turnover of amino acids in organisms. According to biochemistry, an adult man synthesizes and degrades approximately 1 g of protein nitrogen per kg of body weight per day. This corresponds to a protein turnover of about 7.7% per day for a man with a normal body temperature. A too high temperature of the ecosystem (more than about 340 K) will therefore enhance the breakdown processes too much. A temperature range between 260 and 340 K seems, from these considerations, the most appropriate to create the carbon-based life that we know on Earth. An enzymatic reduction of the activation energy makes it possible to realize basic biochemical reactions in this temperature range, without a too high decomposition rate, which would be the case at a higher temperature. In this temperature range anabolic and catabolic processes can, in other words, be in a proper balance.

2.5 NATURAL CONDITIONS FOR LIFE

The conditions for creation of life-ordering processes out of disorder (or more specifically, chemical order by formation of complex organic molecules and organisms from inorganic matter) can now be deduced from the first, second and third laws of thermodynamics:

1. It is necessary that the system be open (or at least non-isolated) to exchange energy (as well as mass) with its environment (see also Chapter 1).
2. An influx of low-entropy energy, that can do work, is necessary (see also Chapter 1).

3. An outflow of high-entropy energy (heat produced by the transformation of work to heat) is necessary (this means that the temperature of the system inevitably must be greater than 2.726 K).
4. Entropy production accompanying the transformation of energy (work) to heat in the system is a necessary cost of maintaining the order.
5. Mass transport processes at a not too low rate are necessary (a prerequisite). This implies that the liquid or gaseous phase must be anticipated. A higher temperature will imply a better mass transfer, and also a higher reaction rate. An increased temperature also means a faster breakdown of macromolecules, and therefore a shift towards catabolism. A temperature approximately in the range of 260–340 K must therefore be anticipated for carbon-based life. See the details in Section 2.4.

The rates of biochemical reactions on the molecular level are determined by the temperature of the system and the exergy supply to the system. Hierarchical organization ensures that the reactions and the exergy available on the molecular level can be utilized on the next level, the cell level, and so on throughout the entire hierarchy: molecules → cells → organs → organisms → populations → ecosystems. The maintenance of each level is dependent on its openness to exchange of energy and matter. The rates in the higher levels are dependent on the sum of many processes on the molecular level. They are furthermore dependent on the slowest processes in the chain: supply of energy and matter to the unit → the metabolic processes → excretion of waste heat and waste material. The first and last of these three steps limit the rates and are determined by the extent of openness, measured by the area available for exchange between the unit and its environment relative to the volume. These considerations are based on allometric principles (Peters, 1983; Straskraba et al., 1997).

In addition to the five conditions given above, it is necessary to add a few biochemically determined conditions. The carbon-based life on Earth requires first of all an abundant presence of water to deliver the two important elements hydrogen and oxygen, as solvent for compounds containing the other needed elements (see below). Water is furthermore a compound that is liquid at a suitable temperature with a suitable diffusion coefficient, has a suitable specific heat capacity to buffer temperature fluctuations and has a suitable vapour pressure to ensure a suitable cycling (purification) rate.

Life on Earth is characterized by about 25 elements. Some of these elements are used by life processes in micro amounts, and it cannot be excluded that other elements could have replaced these elements on other planets somewhere else in the universe. Several metal ions are, for instance, used as coenzymes and are often important parts of high molecular organic complexes. Other ions may be able to play similar roles for biochemical processes and complexes. It is on the other hand difficult to imagine carbon-based life without at least most of the elements used in macro amounts, such as nitrogen for amino acids (proteins—the enzymes) and amino bases, phosphorus for ATP and phosphorous esters in general and sulphur for formation of some of the essential amino acids.

The biochemically determined conditions can therefore be summarized in the following two points:

6. Abundant presence of the unique solvent water is a prerequisite for the formation of life forms similar to the carbon-life forms as we know from the Earth.
7. The presence of nitrogen, phosphorus, sulphur and some metal ions seems absolutely necessary for the formation of carbon-based life.

A last and eighth condition should be mentioned: the seven other conditions should be maintained within reasonable ranges for a very long period of time. The genes may ensure that if an advantageous property of an organism has been developed, the property can be heritaged and the following generations will be able to maintain the advantageous property. The probability to create (complex) life spontaneously is so low that even the time from the big-bang would not have been sufficient. It is therefore necessary that the development towards life is made stepwise with conservation of each achieved progress to allow further development to ride on the shoulders of the already made progress (see Section 2.2). In this context it is important that the presence of life on Earth has changed the conditions for life in a favourable direction. Consider the present oxygen concentration in the atmosphere, which has made it possible to use the more effective biochemical pathways that characterise the aerobic life. Many mechanisms are probably involved in the emergence of a progressive property in the first hand, but indisputable random processes based on trial-and-error are also important in the emergence of progressive properties. This implies that carbon life is not formed overnight. The history of evolution on Earth shows that at a suitable temperature and with abundant water , probably in the order of 10^8 years or more (Haugaard Nielsen, 1999) was needed to form from inorganic components dissolved in water the first living cells with some type of primitive genes to ensure a continuous development (evolution). Fossils after phytoplankton has recently been found at Isua, Greenland, by Minik Rosing (Haugaard Nielsen, 1999). The age of the fossils was determined to be 3.8 billion years old, or about 100 million years after the termination of the massive bombardment of meteors that characterized the first 600–700 million of years after the Earth was born. The focal point is that the seven above-mentioned conditions must be fulfilled for a sufficient long period of time which leads to the eighth condition:

8. As the formation of life from inorganic matter requires very long time, probably in the order of 10^8 years or more, the seven conditions have to be maintained in the right ranges for a very long time, which probably exceeds about 10^8 years.

After the Mars pathfinder mission it has been discussed whether Mars hosts or has hosted life. Clearly, conditions 1–7 are not met on Mars today. The climate is too harsh and water is far from being present in the liquid form and in the amount needed for the planet to bear life. There are, however, many signs of a warmer and wetter climate at an earlier stage. It has been possible to find landscape formations that are similar to the river valleys on the Earth. Fluvial landforms were identified in images of Mars. Recent results from Mars' rovers and orbiters show that warm, wet conditions may have prevailed on the planet for a long period during its early history (Bell, 2006). The methane concentration is also higher at the Mars' equator than closer to the poles, which could be a sign of a bacterial activity. It looks therefore as the seven conditions may have been valid and the question is: "have they also prevailed for a sufficient period of time?". If later missions to Mars will show that it is the case, the next obvious question is: will life inevitably be the result of self-organizing processes, if the eight conditions (the eighth condition about sufficient time, should of course be included) are fulfilled? It should be expected that primitive life had been present on Mars at an early stage, provided that the warmer and wetter conditions had prevailed for sufficient time. The further evolution from unicellular (maybe prokaryotic) organisms to more and more complex organisms as we know from the Earth could not be realized on Mars, because the climate changed and a part of the water and the atmosphere disappeared. Latest investigations of Mars-originated meteorites have made it almost certain that there has previously been microbiological life on Mars. The latest geological investigation (Bell, 2006) has furthermore shown that there has previously been plenty of water on Mars, which also points towards a prior existence of life on Mars. It is of course still an open question if this microbiological life is still present. The Mars pathfinder mission will be able to answer this question.

Another possibility for life in our solar system exists on Europa, one of the moons of Jupiter (Sweinsdottir, 1997). Europa is characterized by a coverage of ice. It implies that there is plenty of water on Europa which means that one of the important conditions for life is fulfilled. In addition, Europa has volcanic activity and the gravity of Jupiter creates heat on Europa. It has been possible in the deep sea to find life close to volcanic activity—the so-called black smokers—although the conditions are very harsh and no light can penetrate more than a few hundred metres through a water column (see more details in Chapter 4). Some researcher (Sweinsdottir, 1997) means that the chance to find life on Europa is higher than on Mars. Europa has of course much less sunlight and the surface temperature is far too low, but the volcanic activity in the deeper parts of the oceans could provide the needed low-entropy energy for the formation and maintenance of life.

Titan, a Saturn moon, has a dense atmosphere of a composition similar to what is estimated was characteristic for the Earth 4 billion years ago. The atmosphere contains even organic compounds (Douthitt, 2006). Locally, higher temperatures from the volcanic-like activities in the deeper layers may have formed the basic molecules for life if not even protocells. Titan could uncover the very first forms of life on the Earth.

Lately, also Enceladus, another of Saturn's moons, became a candidate for primitive life. Enceladus is covered by ice and has a higher temperature than expected, may be due to radioactive processes or may be due to the gravity of Saturn. It has also geysers. Enceladus has water, energy and organic matter, which are the prerequisites for life (Douthitt, 2006).

The hope to find primitive bacteriological life in our solar system has increased due to the rich life of bacteria that has been found in the extreme environment close to the vents and on the subseafloor (Svennevig, 2003). The conditions are anaerobic and often at extreme high temperatures. It is our experience from the Earth that if there is the slighest possibility for life to emerge, it will emerge or be expressed differently: emergence of life seems to be a very robust process.

2.6 THE EVOLUTION AND THE GROWTH OF INFORMATION

Ecosystem development in general is a question of the energy, matter and information flows to and from the ecosystems and between the components of the ecosystems. No transfer of energy in ecological and biological systems is possible without matter and information and no matter can be transferred without energy and information. The higher the levels of information, the higher the utilisation of matter and energy for further development of ecosystems away from the thermodynamic equilibrium (see also Chapter 1). These three factors are intimately intertwined in the fundamental nature of complex adaptive systems such as ecosystems in contrast to physical systems, which most often can be described completely by material and energy relations. Life is therefore both a material and a non-material phenomenon. The self-organisation of life essentially proceeds by exchange of information. Of the three qualities that characterise life—mass, energy and information—it is information that defines life. Indeed, it can be argued that information does not exist in the absence of life (Eigen, 1992). We measure and model mass and energy flows in ecosystems—because we can, but life is information!

The information content increases in the course of ecological development because an ecosystem encompasses an integration of all the modifications that are imposed on the environment. Thus, it is on the background of genetic information that systems develop which allow interaction of information with the environment. Herein lies the importance in the feedback organism—environment, which means that an organism can only evolve in an evolving environment, which itself is modifying. The differences between the two stages include entropy and eco-exergy; (see Chapter 1 for details about these concepts).

The conservation laws of energy and matter set limits to the further development of "pure" energy and matter, while information may be amplified (almost) without limit. Limitation by matter is known from the concept of the limiting factor: growth continues until the element which is the least abundant relative to the needs by the organisms is used up. Very often in developed ecosystems (for instance, an old forest), the limiting elements are found almost entirely in organic compounds in the living organisms, while there is no or very little inorganic forms left in the abiotic part of the ecosystem. The energy input to ecosystems is determined by the solar radiation and many ecosystems capture about 75–80% of the solar radiation which is the upper physical limit. The eco-exergy of the information content of a human being can be calculated by the use of Equations 1.27–1.30 in Chapter 1. The results are presented in the Second Movement: about 40 MJ/g.

A human body of about 80 kg will contain about 2 kg of proteins. If we presume randomly that 0.06 mg of the protein at the most would be different enzymes that control the life processes and therefore contain the most important information, it could represent a maximum content of information. If we presume an average molecular weight of the amino acids making up the enzymes on 200, the amount of amino acid molecules would be $3 \times 10^{-7} \times 6.2 \times 10^{23} =$ about 2×10^{17}. The corresponding β-value can be found from the following equation (see Equation 1.32; Jørgensen et al., 2000, 2005):

$$\beta = 1 + 4.00 \times 10^{-6} \times \text{AMS}$$
$$(\text{AMS} = \text{number of amino acidsin the right sequence})$$
$$= 8 \times 10^{11}$$

It means that 1 g biomass will have the amount of eco-exergy equal to $8 \times 10^{11} \times 18.7 = 1.5 \times 10^{13}$ kJ, as it is presumed that 1 g organic matter has 18.7 kJ of exergy (see Equation 1.30). The calculated amount of eco-exergy corresponds, in other words, to 1.5×10^{13} kJ/g body weight.

These calculations are of course back on the envelop calculations and do not represent what is expected for the information content of organisms in the future; but it seems possible to conclude that the development of the information content is very very far from reaching its limit in contrast to the development of the material and energy relations (see Figure 2.4).

Information has some properties that are very different from mass and energy.

1. *Information can unlike matter and energy disappear without trace.* When a frog dies, the enormous information content of the living frog may still be there a few microseconds after the death in the form of the right amino acid sequences, but the information is useless and after a few days the organic polymer molecules have decomposed.
2. *Information expressed, for instance, as eco-exergy, it means in energy units, is not conserved.* Property 1 is included in this property, but in addition, it should be stressed that living systems are able to multiply information by copying already achieved successful information, which implies that the information survives and thereby gives the organisms additional possibilities to survive. The information is by

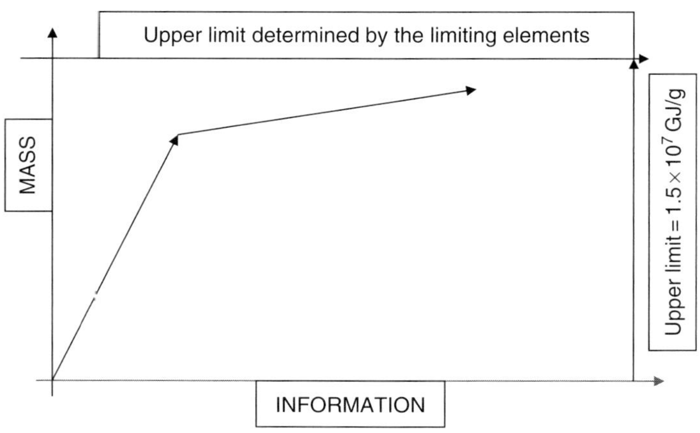

Figure 2.4 Growth of biomass is limited by the limiting elements or by the amount of incoming solar radiation, while the information content of living organisms are very far from its limit.

autocatalysis able to provide a pattern of biochemical processes that ensure survival of the organisms under the prevailing conditions determined by the physical-chemical conditions and the other organisms present in the ecosystem. By the growth and reproduction of organisms the information embodied in the genomes is copied. Growth and reproduction require input of food. If we calculate the eco-exergy of the food as just the above-mentioned average of 18.7 kJ/g, the gain in eco-exergy may be more; but if we include in the energy content of the food, the energy that the evolution has costed, the gain in eco-exergy will be less than the eco-exergy of the food consumed. Another possibility would be to apply emergy instead of energy (see Odum, 1983). Emergy is the cost expressed in the solar energy that is needed to provide a considered product. The emergy of the food would be calculated as the amount of solar energy it has costed to provide the food, which would require multiplication by a weighting factor >>1.

3. *The disappearance and the copying of information, that are characeristic processes for living systems, are irreversible processes.* A made copy cannot be taken back and the life and the death are irreversible process. Although information can be expressed as eco-exergy in energy units, it is not possible to recover chemical energy from information on the molecular level as known from the genomes. It would require a Maxwell's demon that could sort out the molecules and it would therefore violate the second law of thermodynamics. The role of information is directly connected with the problem of Maxwell's demon (Tiezzi, 2006). Brillouin (1962) has shown that the demon would require information about the molecules and that more energy would be needed to obtain this information than the energy gained, which means that Maxwell's demon does not exist. There are, however, challenges to the second law of thermodynamics (see Capek and Sheehan, 2005) and this process of copying information at very low costs could be considered one of them. Mass can be transformed by nuclear processes to energy as expressed by Einstein: $E = mc^2$, but the transformation of energy to mass is not a natural process on the Earth, because it implies that matter and anti-matter are formed, a process that, however, was very common in the early universe. Energy can be transformed to information and macroscopic information to energy (see Figure 2.5); but information on the molecular level would require a Maxwell's demon as mentioned above. Figure 2.6 illustrates the irreversibility of the

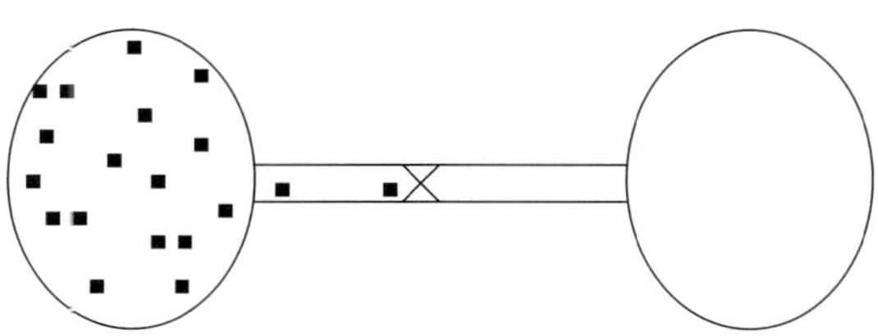

Figure 2.5 The left chamber contains 1 mole of a pure ideal gas, while the right chamber is empty. If we open the valve, the system will loose eco-exergy (or technological exergy) = $RT \ln 2$, which we could utilize by installation of a propeller in the valve. The entropy of the system will simultaneously increase by $R \ln 2$. In this case, macroscopic information is converted to eco-exergy.

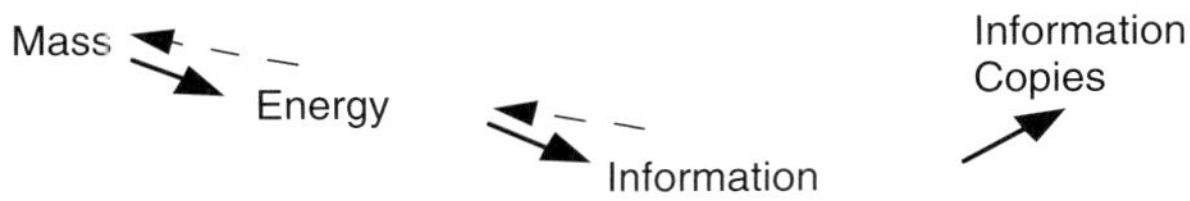

Figure 2.6 Matter can be transformed to energy and energy to information that can be copied at low costs. Transformation of information to energy requires that it is macroscopic information (see Figure 1.5) and transformation of energy to mass implies that both matter and anti-matter are formed. The dotted arrows indicate that the transformation is only possible under certain conditions.

transformation processes between mass, energy and information. It is easy to go from mass to energy to information, but it is not necessarily possible under all circumstances to go from information to energy and from energy to mass.

4. *Exchange of information is communication and it is this that brings about the self-organization of life.* Life is an immense communication process that happens in several hierarchical levels. Exchange of information is possible without participation of mass and energy, while storage of information requires that the information is linked to material, for instance the genetic information stored in the genomes and is transferred to the amino acid sequence.

A summary of the calculations of the information content of organisms and ecosystems are summarised in Box 1.

Box 1 Information

Eco-exergy $= RTcK$ J/l $= 18.7\beta c'$ J/l , where K is Kullback's measure of information, c is the concentration in moles/l and c' the concentration in g/l of the considered components, which could be a living organism. Eco-exergy is, in other words, the product of biomass and the embodied information. 18.7 c' covers the chemical eco-exergy, while $(\beta - 1)\, c'$ is the biological information eco-exergy.

$$\beta = 1 + 4.00 \times 10^{-6} \times \text{AMS}$$

(AMS = number of amino acids in the right sequence)

Information, for instance, expressed as eco-exergy is not conserved.

Biological systems need, in addition to matter and energy, to include the information into a satisfactory description. Living organisms are able to multiply information by copying already achieved successful information, i.e. information giving the organism the property of survival under the prevailing conditions. The multiplication of information is an irreversible process and the information cannot be utilized as an energy source according to an extended version of the Second Law of Thermodynamics, but the information in living organisms is able, frequently by autocatalysis, to provide an effective pattern of biochemical processes that ensure the survival of the organism under the prevailing conditions, determined by the physical-chemical conditions and the other organisms present in the ecosystem (or expressed differently, contribute to the synergy of the ecological network). The information is lost by the death.

2.7 IS LIFE A MIRACLE?

How can we explain the diversity of life—the biodiversity, which is rooted in the enormous variability of the life conditions in time and space? How can we explain the beauties of nature: the colour and pattern of the butterflies, the spectacular colour symphony of a temperate forest at fall, the songs of the birds at dawn, the bright and soft skin of a polar bear and many more examples?

All known life-forms on Earth reside in the thin layer enveloping the globe known as the ecosphere. This region extends from sea level about 10 km into the ocean depths and approximately the same distance up into the atmosphere. It is so thin that if an apple were enlarged to the size of the Earth, the ecosphere would be thinner than the peel. Yet a vast and complex biodiversity has arisen in this region.

However, even in this limited domain the conditions for living organisms may vary enormously in time and space.

The climatic conditions:

1. The temperature can vary from about −70 to about +55 °C.
2. The wind speed can vary from 0 km/hr to several hundred km/hr.
3. The humidity may vary from almost 0 to 100%.
4. The precipitation can vary from a few mm in average per year to several mm per year which may or may not be seasonally aligned.
5. Annual variation in day length according to longitude.
6. Unpredictable extreme events such as tornadoes, hurricanes, earthquakes, and volcanoes.

The physical-chemical environmental conditions:

1. Nutrient concentrations (C, P, N, S, Si, etc.)
2. Salt concentrations (it is important both for terrestrial and aquatic ecosystems)
3. Presence or absence of toxic compounds, whether they are natural or anthropogenic in origin
4. Rate of currents in aquatic ecosystems and hydraulic conductivity for soil
5. Space requirements.

The biological conditions:

1. The presence of food for herbivores, omnivores and carnivores organisms
2. The presence of predators
3. The presence of competitors for the resources (food, space, etc.)
4. The presence of pollinators, symbiants and mutualists
5. The presence of decomposers.

The human impact on natural ecosystems today adds to this complexity.

The list of factors determining the life conditions is much longer—we have only mentioned the most important factors. In addition, the ecosystems have history or path dependency meaning that the initial conditions determine the possibilities of development. If we modestly assume that 100 factors are defining the life conditions and each of these 100 factors may be on 100 different levels, then 10^{200} different life conditions are possible, which can be compared with the number of elementary particles in the universe as 10^{81}. The variability is unconceivable. The confluence of path dependency and an astronomical number of combinations affirm that the ecosphere could not experience the entire range of possible states, otherwise known as non-ergodicity. Furthermore, its irreversibility ensures that it cannot track back to other possible configurations. In addition to these combinations, the formation of ecological networks means that the number of indirect effects is magnitudes higher than the direct ones and they are not negligible—on the contrary, they are often more significant than the direct ones.

What is the result of this enormous variability in the natural life conditions? We have found about 0.5×10^7 species on Earth and it is presumed that the number of species is twice as much or 10^7. They have developed all types of mechanisms to live under the most varied life conditions including the ones at the margin of their physiological limits. They have developed defence mechanisms—plants are toxic to avoid grazing, or have thorns and so on; or herbivores animals have developed horns, camouflage pattern, well-developed auditory sense, fast escaping rate and so on. They have furthermore developed integration mechanisms; fitting into their local web of life, often complementing and creating their environmental niche. The multiplicity of the life forms is inconceivable.

The number of species may be 10^7, but all living organisms are different from each other. An ecosystem has normally 10^{15} to 10^{20} individual organisms that are all different, which although it is a lot, make ecosystems middle number systems. This means that the number of organisms is magnitudes less than the number of atoms in a room—but all the organisms, opposite the atoms in the rooms, have individual characteristics. Whereas large number problems such as the number of atoms in a room are amenable to statistical mechanics and small number problems such as planetary systems to classical mechanics or individual-based modelling, middle number problems contain their own set of challenges. For one thing this variation, within and among species, provides diversity through co-adaptation and co-evolution, which is central to both Darwinian selection and network aggradation.

The competitive exclusion principle (Gause, 1934) claims that when two or more species are competing about the same limited resource, only the best one will survive. The contrast between this principle and the high number of species has, for long time, been a paradox. The explanation is rooted in the enormous variability in time and space of the conditions and in the variability of a wide spectrum of species' properties. A competition model, where three or more resources are limiting, gives a result that is very different from cases where one or two resources are limiting. Due to significant fluctuations in the different resources, it can be shown by use of a competition model that many species competing about the same (wide) spectrum of resources can coexist. It is therefore not surprising that there exist many species in an environment characterized by an enormous variation of both abiotic and biotic factors.

To summarise, the number of different life forms is enormous because there are a great number of both challenges and opportunities. The complexity of ecosystem dynamics is rooted in these two incomprehensible types of variability.

Another important factor that explains the magnificent result of the evolution is the time. The very first life emerged about 4 billion years ago. Bacteria have a doubling time of a few hours and the generation time of *Homo sapiens* is about 25 years. Unicellular organisms with a short generation time have dominated the Earth in the first billion years. If we presume an average generation time of one day for the entire evolution (the first 3 billion years of the evolution the generation time has probably been shorter) which is of course a very rough estimate, the entire evolution corresponds to about 10^{12} generations. From many studies, we know that a small advantage in the fight for survival is crucial and may determine the evolutionary direction. So, if small advantages are added up generation by generation, it is possible to explain the evolution from prokaryotic cells to humans. If we presume an advantage (change) of 1% per generation, it yields a change of about 10^{12}% in 10^{12} generations! As already underlined above, the number 10^{12} is an extremely rough estimate and it will never be possible to assess the right number of generations, but the calculations are not made to find a right number of possible changes, but just to demonstrate that 4 billion years is a very long time, which most probably is sufficient to explain the enormous difference between a prokaryotic cell and *Homo sapiens.* The number of mutations is in the order of $1{:}10^6$ and with the enormous amount of organisms on Earth, the 1% advantage per generation is probably at least in some periods on the lower side of what is possible (see "A short history of nearly everything by Bill Bryson", 2003). Add on top of this the enormous variability of life conditions in time, and it becomes clear that the many life forms and their beauty can easily be explained as the result of natural processes. The almost infinite variety of the genes provides for changes in a changing world. Life is not a miracle, but a result of natural processes. Hopefully, the readers will be convinced when they have had occasion also to read the subsequent chapters. Life as we know it on the Earth is a natural consequence of the second law of thermodynamics. The prerequisite are:

- the irreversibility principle as mentioned above to "create" the directionality
- a constant input of energy to be used to move the system away from thermodynamic equilibrium
- the right life conditions with respect to temperature
- concentrations of nutrients
- the variability of the life conditions
- sufficient time.

3

The evolutionary mechanisms

3.1 INTRODUCTION

In Chapter 2, we have discussed the interplay between constraints and evolution. Without constraints there is no evolution; in other words without the right constraints there is no evolution, because the carbon-based life can only evolve in the right range of temperature and concentrations of a certain number of elements, as discussed in Chapter 2. New constraints that are introduced in the right time are, however, able to increase the rate of the evolution. There is a strong selection pressure of course on the properties that are able to meet the new constraints. The fitness to the new prevailing conditions resulting from the new constraints may be low in the beginning, but due to a selection pressure concentrated on the property or the few properties that are decisive for new constraints, the fitness will relatively rapidly increase or, expressed more correctly, the non-fitted organisms will rapidly be eliminated making the surviving population better fitted and with genes that ensure that the next generation is better fitted. Those individuals who are most efficient in coping with the challenges of the environment and in competing with other members of their populations and with those of the other species will have the best chance to survive until the age of reproduction to reproduce successfully (Mayr, 2001).

No individual is perfectly adapted, because every genotype represents a compromise of genetic variability and stability. Environments are perpetually changing and at the end of a drought period a population will be better adapted for the drought conditions than for a wet period; see below the example of a structurally dynamic model for Darwin Finches in a drought period. The genotype strikes, in other words, a balance between conflicting demands.

Figures 3.1 and 3.2 illustrate the difference between changing constraints and new constraints with a stronger selection pressure on one or a few properties, which may correspond to a few genes. Notice, however, that it is always the phenotype that is selected, while the information about the well-fitted properties of the phenotypes are in the genes.

Five to six million years ago the climatic changes turned rain forest into a tree savanna stage. Fortunately, the chimpanzee-like apes at that time had knuckle walk and were from time to time even bipedal. So, there was a relatively good starting point for the development of bipedal apes, which would be better fitted to the savanna. The new line of evolution, *Australopithecus*, developed very fast as a result of new constraints due to climatic changes at the right time. The property in focus and selected for

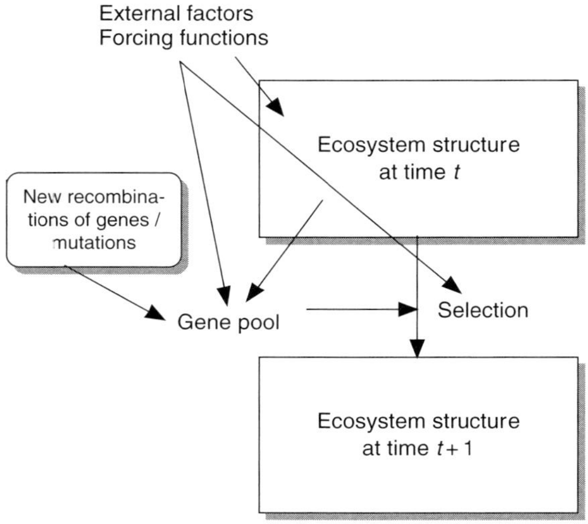

Figure 3.1 Conceptualization of how the external factors (constraints) steadily change the ecosystem structure, i.e., the species composition and their properties. The possible shifts in species composition and their properties are determined by the gene pool, which is steadily changed due to mutations and new sexual recombinations of genes. The development is, however, more complex, as will be discussed in the next chapter. The selection pressure is presumed to be on a high number of phenotypes and thereby on a wide spectrum of properties. The selection pressure implies an elimination of the species and the individuals with properties that are not well fitted to the currently changing conditions.

was the ability to use bipedal locomotion, which was beneficial compared with knuckle walk. There was no need to change most of the ancestral chimpanzee characters, such as small brain, long arms, short legs and large sexual dimorphism.

The development of the human brain is another example of new constraints at the right time will imply a relatively faster evolution. *Australopithecus* was bipedal for more than 2 million years without any significant change in the brain volume. They did not produce flaked stone tools. Beginning about 2.5 million years ago, the climate began, however, to deteriorate. It became more arid and the trees in the savanna suffered and died. The savanna changed to a bush savanna. This deprived the australopithecines of their retreat to safety. Compared with a tree savanna they were completely defenseless. They were threatened by lions, leopards, hyenas and wild dogs, all of whom could run faster than them. Some of the australopithecines survived by inventing successful defense mechanism. The survivors could have thrown rocks or used weapons made from wood. They may have had long poles, swung thorn branches and used noise-making instruments. They were probably also the first humans to make flaked stone tools. *Homo habilis* who lived about 2 million years ago was also in the need – maybe in an urgent need to provide more meat. The nourishment encompassed carcasses, vegetables and fruit. *Homo habilis* had primitive

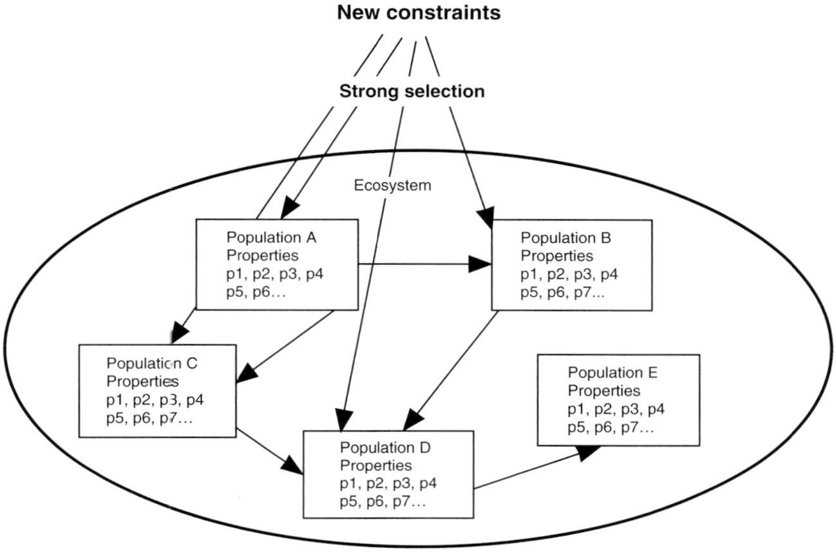

Figure 3.2 New constraints may imply that the selection pressure is concentrated on a few specific properties of a few populations in the ecosystem. These properties will be changed relatively rapidly due to the strong selection pressure on the few genes that determine the properties. In the figure, new constraints impose a strong selection of property p5 for populations *A*, *B* and *D* and of property p2 for population *A*. Notice that the selection is taken place on the phenotype level.

flint stone tools to facilitate the collection of food. It was therefore natural to try to hunt animals by a cooperative effort of the tribe and develop better tools to facilitate also hunting. A new constraint had emerged on *Homo habilis:* to be a better planner and a better tool maker, both activities requiring bigger brains. This shift was the most fundamental one in all of the hominid history. It was far greater than the change to bipedal locomotion.

The brain size increased over a period of 2 million years from 650 cm^2 for *Homo habilis* to 1350 cm^2 for *Homo sapiens*. If we consider one generation to be 20 years, the period corresponds to 100,000 generations.

How much should the brain grow per generation to make more than a doubling possible during 2 million years? A simple calculation shows that the brain has to increase less than 0.001% per generation, which however still is considered relatively fast. This example also shows that a very long time has been available for the entire evolution, and that the stepwise, slow evolution accumulating for every step small advantages can explain the development towards higher and higher complexity.

We mention a third example to illustrate how new constraints entail a relatively fast evolution due to a specific selection pressure. The fast increase in the beak size of Darwin's finches has been modelled by Jørgensen and Fath (2004).

The models reflect therefore—as all models—the available knowledge, which in this case is comprehensive and sufficient to validate even the ability of the model to describe the changes in the beak size as a result of climatic changes, causing changes in the amount, availability, and quality of the seeds that make up the main food item for the finches. The medium ground-finches, *Geospiza fortis*, on the island Daphne Major were selected for this modelling case due to very detailed case specific information found in Grant (1986). The model has three state variables: seed, Darwin's finches adult and Darwin's finches juvenile.The juvenile finches are promoted to adult finches 120 days after birth. The mortality of the adult finches is expressed as a normal mortality rate (Grant, 1986) plus an additional mortality rate due to food shortage and also that caused by a disagreement between bill depth and the size and hardness of seeds.

The beak depth can vary between 3.5 and 10.3 cm (Grant, 1986) and the beak size is equal to $\sqrt{DH}$, where D is the seed size and H the seed hardness that are both dependent on the precipitation, particularly during the months January–April (Grant, 1986). It is possible to determine a handling time for the finches for a given $\sqrt{DH}$ as function of the bill depth (Grant, 1986), which explains that the accordance between $\sqrt{DH}$ and the beak depth becomes an important survival factor. The relationship is used in the model to find a function called "diet", which is compared with $\sqrt{DH}$ to find how well the bill depth fits to the $\sqrt{DH}$ of the seed. This fitness function is based on the information given by Grant (1986) about the handling time. It influences, as mentioned above, not only the mortality of adult finches but also has an impact on the number of eggs laid and the mortality of the juvenile finches. The growth rate and mortality of seeds are dependent on the precipitation which is a forcing function know as function of time (Grant, 1986).

A function called shortage of food is calculated from the food required for the finches which is known (Grant, 1986), and the available food (=the seed state variable). How the food shortage influences the mortality of juvenile finches and adult finches can be found in Grant (1986). The seed biomass and the number of *G. fortis* as a function of time from 1975 to 1982 are known (Grant, 1986). These numbers from 1975 to 1976 have been used to calibrate the following parameters:

i. The influence of the fitness function on (a) the mortality of adult finches, (b) the mortality of juvenile finches and (c) the number of eggs laid.
ii. The influence of food shortage on the mortality of adult and juvenile finches is known (Grant, 1986). The influence is therefore calibrated within a narrow range of values.
iii. The influence of precipitation on the seed biomass (growth and mortality).

All other parameters are known from the literature.

The eco-exergy density is calculated (estimated) as $275 \times$ the concentration of seed $+980 \times$ the concentration of Darwin's finches (see Table 1.1). Every 15 days it is found if a feasible change in the beak size, taking the generation time and the variations in the beak size into consideration, will give a higher eco-exergy. If it is the case, then the beak size is changed accordingly. The modelled changes in the beak size were confirmed by the observations.

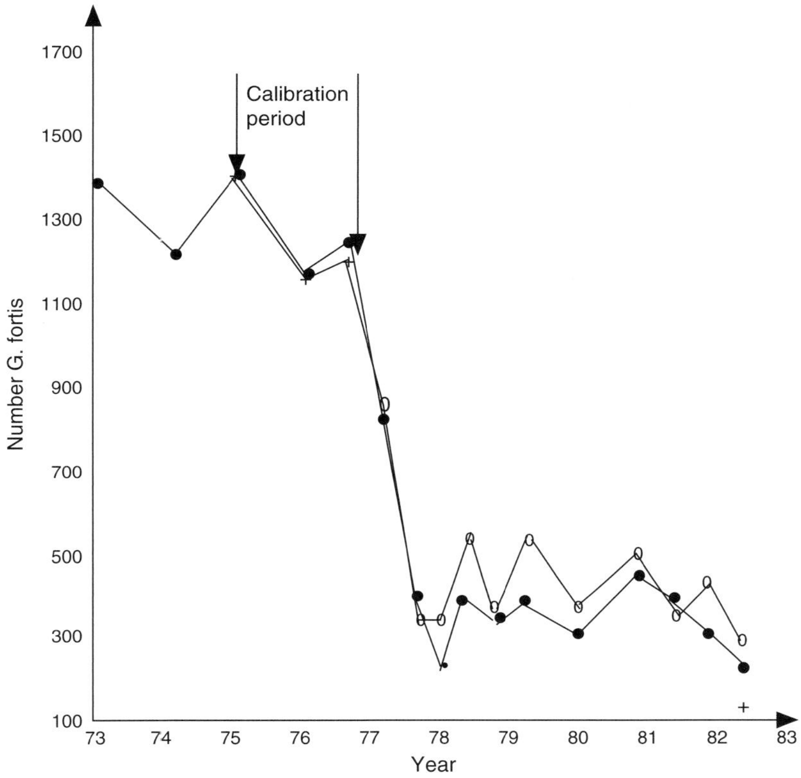

Figure 3.3 The observed number of finches (•). From 1973 to 1983, compared with the simulated result (0): 75 and 76 were used for calibration and 77/78 for the validation.

The model results of the number of Darwin's finches are compared with the observations (Grant, 1986) in Figure 3.3. The standard deviation between modelled and observed values was 11.6% and the correlation coefficient, r^2, for modelled versus observed values was 0.977. The results of a nonstructural dynamic model would *not* be able to predict the changes in the beak size and would therefore give too low values for the number of Darwin's finches, because their beaks would not adapt to the lower precipitation yielding harder and bigger seeds.

3.2 FOUR INHERITANCE SYSTEMS

The role of the genes in the evolution has also been emphasized in Chapter 2. Without the genes, or rather a heritage system, there would have been no evolution or at least it would have been at a much slower rate. The interplay between the three growth forms has also a role, because the growth of biomass and networks may create constraints, which inevitably will influence the third growth form from the information that is embodied in the genes.

The neo-Darwinian theory has taught us that the adaptation occurs through natural selection of changed genetic variations. It is a part of the story, but not the full story. Let us denote it as heredity system number 1. It includes the so-called *Hox* genes, which play a pivotal role in specifying regional identity in body plans. It has been suggested that increasing complexity of body plans during evolution might be causally correlated with increasing complexity of the Hox gene complexes (Mayr, 2001).

It has been shown that cells can transmit information to daughter cells through non-DNA (epigenetic) inheritance. Let us call it heredity system number 2. In addition, many animals transmit to others by behavioural means, which may be considered as a heredity system 3. For instance, the bear mother shows the cubs by her behaviour which food is best to eat and how to catch salmons. Finally, we have the symbol-based heredity system, particularly language, which has played an increasingly more important role throughout the evolution of the heredity system. The bear mother can also teach the cubs about dangers by sounds. The language of course plays an enormous role for *Homo sapiens* as a heredity system. The four different heredity systems will be presented in more detail below. We will discuss how the presence of the four heredity systems is able to explain why we have observed at least in some phases a very, sometimes surprisingly, rapid evolution. When all four inheritance systems and the interactions between them are taken into account, we get a different view of the Darwinian evolution (Jablonka and Lamb, 2006). The evolution is under all circumstances more complex than we thought a few decades ago. The evolution cannot be explained by one straightforward mechanism, but the right explanation is rooted in possible interactions of several different mechanisms.

3.3 THE DARWINISM AND NEO-DARWINISM

Darwin's theory is built on four principles or properties:

1. *The reproduction.* The organisms are able to reproduce and every population has such high fertility that its size would increase exponentially if not constrained.
2. *The inheritance.* The properties of the offsprings are almost inherited unchanged from the parents or from the mother cell. This is the basis of the heredity system number 1, the genes. As pointed out in Chapter 2, the evolution would be impossible without the possibilities of building on the shoulders of what has already been achieved. The change of the genes may be random—by mutation and sexual re-combinations—but once a good gene has been found, i.e. genes that give a high probability of survival and further growth, the information will be transferred to the next generation due to the genes.
3. *The variation.* Not all organisms are identical but they are all more or less different. The variation is needed to have a selection. If all the organisms were identical, there would be no differences in survival and growth and therefore no selection. The genes are different from organism to organism, even among organisms of the same species. So, the selection takes place among the differences in properties of the phenotypes, resulting in currently changing genomes generation after generation. A selection due,

for instance, to the introduction of new constraints will imply that the distribution in the variation of a property will move in a direction of what yields the best fitness (see Figure 3.4). Later, principle 2 presented above may cause more pronounced changes of the properties with even better fitness.
4. The competition, i.e. the resources are limited. Therefore, only some of the organisms can survive. The other will be outcompeted or eliminated in the fight about the resources. The heritable variation affects the success of the organisms in surviving and multiplying.

The variation, principle 3, represents the chance in the Monod sense: evolution is a trade off between chance and necessity (Monod, 1972), and the selection, the reproduction and the heritage (principles 1 and 2) as a result of limited resources (principle 4), represents the necessity.

Lamarck's theory is often associated with the idea that acquired characters are inherited. We know that it is wrong. Lamarck's theory encompasses much more than the inheritance of acquired characters. Moreover, Lamarck did not invent the theory that acquired characters are inherited, because all biologists believed this at the time of Lamarck. Darwin's theory or rather the Neo-Darwinism, which includes Mendel's heredity laws, explains the heredity system number 1. Lamarck, however, did not explain but discussed briefly on the heredity systems 3 and 4.

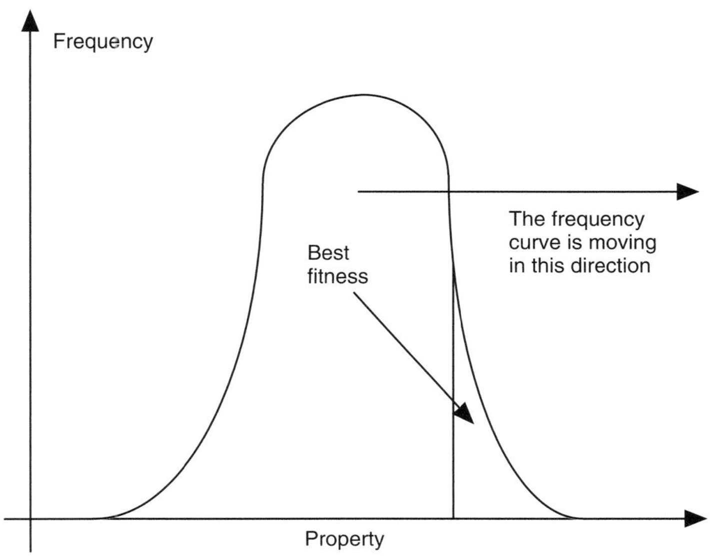

Figure 3.4 A selection due, for instance, to the introduction of new constraints will firstly imply that the distribution in the variation of a property will move in a direction of what yields the best fitness.

This book focuses mainly on the thermodynamic interpretation of the evolution and we will therefore not go into details about neo-Darwinism and the many objections against the most rigid form of neo-Darwinism, but will mention Dawkins's ideas. They were published in 1976 in his book *The Selfish Gene*. According to Dawkins, heredity and variation cannot be influenced by adaptive processes that go on in individuals. Dawkins' theory about the selfish gene has of course been criticized. Although the two views that either the genes or individuals and groups are the focus of the selection are not incompatible, there was fierce fighting about the scientific truth in the 1970s and the early 1980s. The discussion was on the following

- The units of variation, DNA sequence or the characters
- The origin of variation, random mutations and sexual recombinations or random DNA changes
- Target of selection, individuals, cells, organs and groups or the genes
- Unit of evolution, the population of individuals or the population of alleles of the gene.

In this context, it is necessary to distinguish between the genotype and the phenotype. The genotype is an organism's inherited potential. Whether this potential is realized is a question about the environmental conditions. The genotype plus the environmental conditions determine the phenotype, which is also the level of selection.

The Human Genome Project delivered the first draft of the promised sequence of human DNA in 2001. We have about 35,000 genes. How can these relatively few genes explain the properties and the variability of humans and the events that occur during the embryonic and postembryonic development. The answer can be found in the properties of the other heredity systems.

The genes are currently changed using several mechanisms:

1. The genes are changed due to mutations, which are not totally unregulated processes, i.e. the mutations are not entirely random, but occur somewhere between random and directed (Jablonka and Lamb, 2006).
2. The genes are changed due to sexual recombinations. Preexisting gene variations are shuffled to produce new combinations. This is the ultimate genetic reason for sex, namely the potential for creating new and better gene arrangements that can advance evolution through recombination.
3. The cells have enzymes that can change RNA that is transcribed from DNA. They have also enzymes that can cut, splice, and mess around with DNA itself. A normal chromosome can be deleted, amplified and rearranged.
4. McClinktock (Futuyma, 1986) found by working with maize that genes on chromosomes actually move around and transpose themselves and were even changed by environmental stress factors.
5. Polycellular organisms are the results of symbiotic relationships among many unicellular organisms (Margulis, 1981). It may generally explain jumps in evolution; two or more properties are suddenly united and create a symbiotic effect.

6. Bacteria can create variations in the genes by conjugation, which means that they exchange genes, for instance through a thin tube. They can also pick up genes which they find in the environment. Conjugation makes it possible for bacteria to approach the advantages that are characteristic for the sexual recombination.

3.4 THE EPIGENETIC INHERITANCE SYSTEM

DNA is not solely responsible for all the hereditary differences between individuals. As already mentioned, the observed differences require that we include three more heredity systems.

The different cells in a polycellular organism look different, behave differently and function differently; for instance, in our body the kidney cells, the liver cells, the skin cells and the cells of the muscles. The differences are not entirely genetic but epigenetic. They are consequences of events that occurred during the development of each type of cell and determine which genes are turned on and how their products act and interact. It is remarkable that many specialized cells can maintain their own particular phenotype for long periods and even transmit it to daughter cells. When kidney cells divide they are kidney cells. Cells acquire information that they can pass to their progeny. The information is transmitted through what is known as the epigenetic inheritance system (EIS). This is the second dimension of heredity and it has been a very powerful mechanism in the evolution. It may also be denoted as cell memory because it remains in a cell and in its descendants for life. It is the deposit of information for all cellular activities.

Before the mid-1970s, the biologists were mostly concerned with the signals that switch the genes on and off. The emphasis was on how cells acquire their specialized role rather than on the problem of how, once the appropriate genes have been turned on and off, the cells are able to remember their new epigenetic state and transmit it to the progeny. Today, biologists have gained a lot of knowledge about EIS, but they tend to associate them with ontogeny—with the process through which the fertilized egg develops into an adult organism with specialized cells and organs. The emphasis is on the determination and regulation of cellular activities, while there is also an enormous evolutionary potential in EIS. We will mention below four broad categories of EIS. They all have a role in cell heredity and in the evolutionary potential.

The first type of EIS is that daughter cells can inherit patterns of gene activity present in the parent cell. They do so when the control of gene activity involves self-sustaining feedback loops. Figure 3.5 shows how the system works. When a gene is active, a protein is produced, which among others acts as a regulator attaching itself to the control region of the gene, whereby it is kept active long after the original inducing cue has disappeared. If the level of the protein remains high enough in each daughter cell, it will continue to act as a positive regulator and the gene will remain active in both cells.

The second type of EIS is concerned with the cell structures. Alternative versions of some cellular structures can be inherited because existing structures guide the formation of similar ones in daughter cells. There are many examples of this type of EIS but we will quote one taken up by Cavalier–Smith. He has considered how the many types of membrane in an ordinary cell are formed. Cell membranes differ from each other in

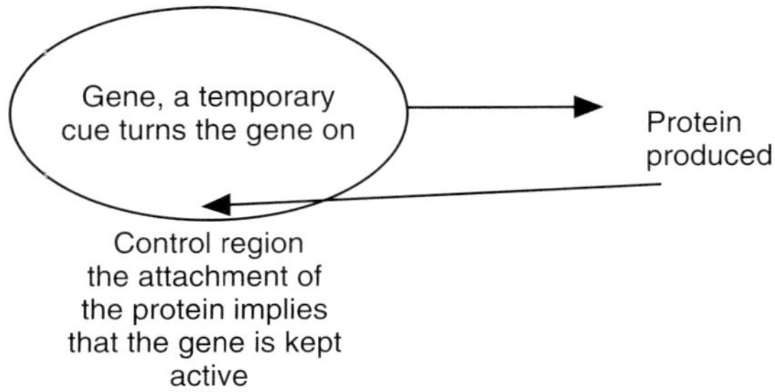

Figure 3.5 When a gene is active, a protein is produced, which among other things acts as a regulator attaching itself to the control region of the gene, whereby it is kept active long after the original inducing cue has disappeared. If the level of the protein remains high enough in each daughter cell, it will continue to act as a positive regulator and the gene will remain active in both cells.

composition and in location. Such membranes cannot assemble without guidance/ "instructions". The preexisting membranes template the formation of more membranes with the same instructions. Through this template, the membranes grow and are eventually divided between daughter cells. They may be called perpetuating membranes of the cells, because like the genome they carry hereditary information in its structure.

The third type of EIS is called chromatin-marking system. Chromatin is the stuff of chromosomes, i.e. the DNA plus the RNA, proteins and other molecules associated with it. The small proteins in eukaryotes, called "histones", are a necessary part of chromosomes. They play a role in compacting the DNA. Slightly less than two turns of DNA is wound around a core of eight histones to form a beadlike structure known as a nucleosome. The non-DNA features of chromatin are transmitted from generation to generation and enable states of gene activity or inactivity to be perpetuated in cell lineages. Methylated DNA is found in all vertebrates, plants and in many invertebrates. A methyl group is attached to the base cytosine. It does not change the codes for a protein, but influence the likelihood that it will be transcribed. Methylation patterns influence how easily the genes can be turned on and off and they are a part of the heredity system that transfer epigenetic information from mother cells to daughter cells. Methylation patterns can be reproduced because they participate in the semi-conservative replication of DNA.

The fourth type of EIS is denoted as RNA interference. It opens up fantastic opportunities for manipulating cells, combating diseases and engineering new qualities into organisms. It was discovered by some gene manipulation experiments that the genes of interest became silent. Large mRNA can be chopped into small pieces, 21–23 nucleotides long. These pieces can cause the destruction of copies of the abnormal mRNA, from which they have been derived. They do so by base paring with the complementary sequence in the mRNA and thereby guide another enzyme to degrade the molecule. The RNA interference system seems to defend cells against invading viruses and the activities of genomic parasites. The idea that RNA interference is a cellular immune system makes sense of many of the odd properties, such as the small pieces of mRNA, mentioned above can amplify and has the ability to make genes silent.

It cannot be excluded that we will find more EIS types in the future, but it is clear that the four types encompass a heredity system beyond the DNA.

The magnitude of the indirect effects of EISs on evolution has been enormous. This is obvious if we think about complex organisms with cells specialized to do different jobs. Without the cell memory, plants and animals with many differentiated cells could not have evolved. EISs, which provide cell memory and enable cell lineage to maintain their characteristics., are the preconditions for the evolution of more and more complex organisms.

During the Cambrian Explosion a shift to the supracellular information of the body plan took place (Barbieri, 2003). A longer embryonic development became possible and could produce more complex animals. It is characteristic that the body plan already existed. All that was needed was a transformation of that potential information into actual information (Barbieri, 2003). The Cambrian Explosion was a transition from a primitive type of development that was totally controlled by genes to a discontinuous type of embryonic development that could also use a body plan.

3.5 THE BEHAVIOURAL INHERITANCE SYSTEM

Mammals, birds and other animals as well can learn from their experience. What is the importance of such learning in evolution?

A lion mother is an excellent hunter and therefore she is able to feed many offsprings and distribute her hunting genes, covering her ability to make a fast hunting tactic and to run rapidly in the right moment, to them. Her lion cubs will most probably survive because they obtain sufficient food and they will to a high extent inherit her excellent hunting genes. In addition, she can teach them her hunting strategy and will have more time to care for them in general due to her successful hunting. So, the cubs not only survive and it means that the genes also survive, but also the better nursing and the better hunting strategy survive through the learning/behaviour from one generation to next. In accordance with the glossary of the computer age, we can say that not only the hardware—the genes—but also the software—the know-how—survives. It is interesting in this context that the heritage of the good hardware will enhance the possibilities of the software to "survive".

There are numerous examples of the role of socially mediated learning and how it implies that new habits, skills and preferences are transferred from generation to generation.

3.6 THE SYMBOLIC INHERITANCE SYSTEM

The use of communication based on symbols is the fourth heredity system. Many animals have, for instance, a simple language that makes it possible to communicate among individuals in a population. The application of the symbolic inheritance system has of course increased enormously during the last few million years due to development of a more and more sophisticated language for the various homo species. The development of the written language, later the art of printing and recently the internet are all major steps in the further development of the symbolic inheritance system. The evolution based on the cultural development is going at a much higher rate than the first heredity system. We may go so far as to express that the evolution has almost entirely been taken over by the fourth heredity system.

3.7 THE SEMANTIC EVOLUTION

The evolution of any organic code is an important evolutionary step. It is presumed that the origin of a code corresponds to the appearance of a complete set of rules. When that happens, something totally new appears in nature, something that did not exist before. According to semantic biology, several organic codes exist in life. Their appearance in Earth marked a great historical event. The semantic biology (see Figure 3.6) operates with at least five codes: the genetic code, the splicing code, the adhesion code, the pattern code and the linguistic code.

Splicing is the cutting-and-pasting operation, for instance by removing some RNA strings and joining together the remaining pieces. It is suggested that splicing is

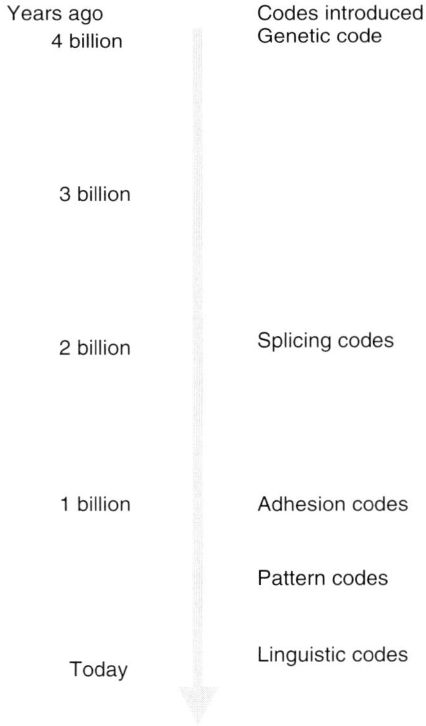

Figure 3.6 The semantic biology operates with at least five codes: the genetic code, the splicing code, the adhesion code, the pattern code and the linguistic code.

a codified process because it is implemented by structures that are very similar to those of proteins. The splicing catalyst is know as spliceosomes. They are huge molecular machines with a molecular weight in the range of ribosomes. For further detail see Barbieri (2003). The adhesion codes are covered mainly by the heredity system 2, while the pattern codes cover the spatial organisation in the embryonic development. An arm and a leg are made of the same tissues and yet their shapes are different, which means that during limb development the same cells can be arranged in different spatial configurations. Pattern genes are a wide class of genes that control the entire body plan of the organisms—a kind of spatial master plan. The linguistic code is obviously the same as heredity system 4—the use of symbols.

The actual number of organic code is likely to be much higher than the five codes presented in Figure 3.6. The emergence of the evolutionary mechanisms are described with reference to the book *The Organic Codes* by Barbieri (2003).

3.8 ILLUSTRATION OF THE ROLES OF THE VARIOUS EVOLUTIONARY MECHANISMS

In this section, model illustrations of some of the evolutionary mechanisms are presented. This section will focus on the role of

1. sexual recombinations offering currently better and better solutions by providing constantly new gene combinations by shuffling the existing genes,
2. random emergence of new modifications of the genes by mutations,
3. the learning process – the heredity system 3, and
4. combining two or all three of these mechanisms.

The models are developed in STELLA and are all very simple. They are giving a first *coarse* illustration of the mechanisms and their differences. It cannot be excluded of course that other models of the mechanisms in the future will give more details and be more realistic. The model results cannot be considered quantitatively, but in the best case only semi-quantitatively. It is, however, assumed that the combination of two or all three mechanisms in a model will give some first-hand indication of the acceleration of the evolution that can be expected when more evolutionary mechanisms are working together. The illustration of the effect on the rate of evolution of combining two or more evolutionary mechanisms was considered a major motivation for the development of the models shown in Figures 3.7 and 3.8.

Figures 3.7 and 3.8 show the eight models that have been applied for the illustration of the three mechanisms and their combinations. The state variables $A \rightarrow H$ represent all the eco-exergy. The first model, A, is just a reference. No increase in eco-exergy is recorded because the input is not increased and the output is equal to the input. Model A represents a steady-state situation. The second model, B, represents a constant increasing input due to new possibilities to find new slightly better solutions by new sexual recombinations. The third model in Figure 3.7 represents the mutations by a random input varying from 0 to 0.015. The mutations are often negative, but the worse solutions

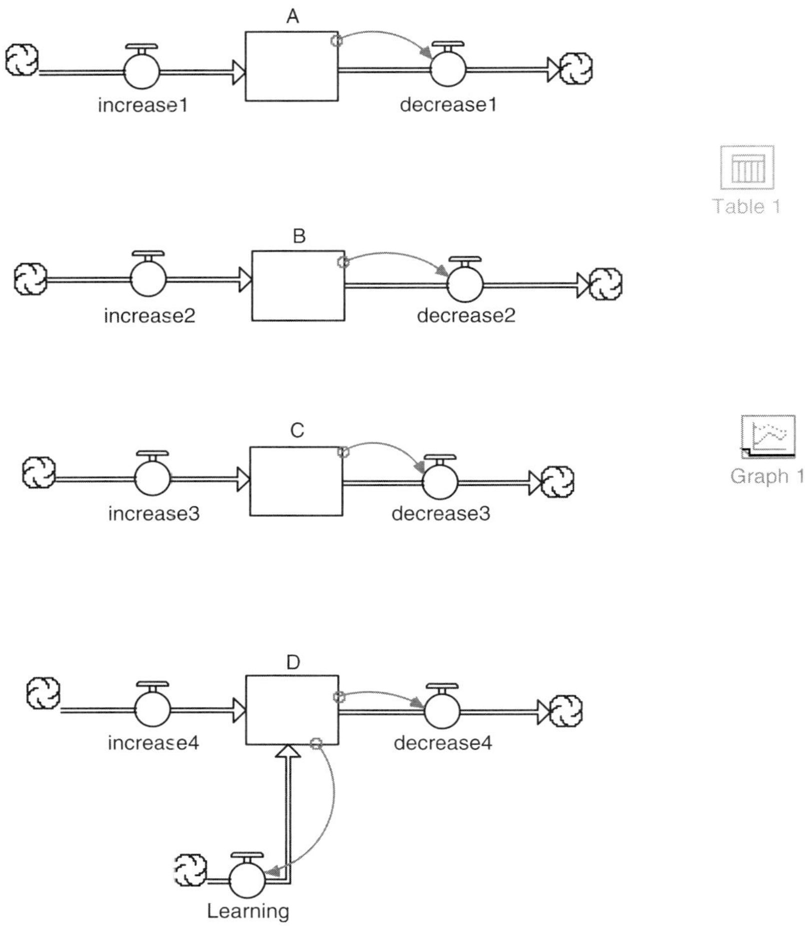

Figure 3.7 Four models representing different evolutionary mechanisms are shown. The first model with state variable *A* represents steady state: no changes in eco-exergy. The second model represents a constant increasing input due to new possibilities to find new slightly better solutions by a steady emergence of new sexual recombinations. The third model, with the state variable *C*, illustrates mutations. The last model, with the state variable *D*, represents the evolutionary mechanism, learning.

are eliminated by the selection pressure and therefore need not be included. The fourth model considered the learning. If the eco-exergy is more than a selected value, the "lion, tiger or bear mother" will have some time to teach the cubs to hunt, because the little family has sufficient food due to the hunting success. If, on the other hand, she is not a good hunter, she is fully occupied to provide food for the cubs and herself to attempt to survive and there is no time available to teach the cubs how to hunt. The more the

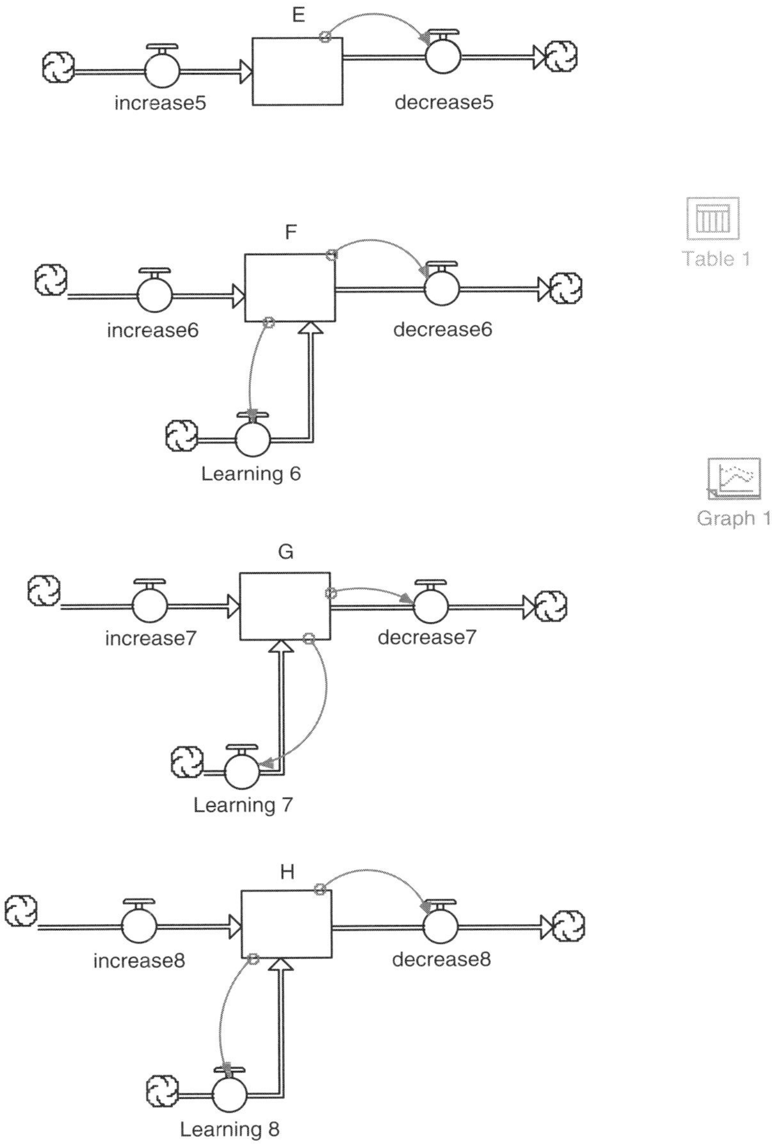

Figure 3.8 Models *E* to *H* are used to illustrate the combinations of the three evolutionary mechanisms shown in Figure 3.7. The fifth model based on the state variable *E* combines sexual recombinations with mutations. The sixth model with *F* as state variable combines the sexual recombinations with learning, and the seventh model based on the state variable *G* combines the mutations with learning. The final model, combines all three mechanisms: sexual recombinations, mutations and learning.

eco-exergy increases the more time she will have and therefore the better the cubs learn how to hunt from their mother. It gives them some advantages that they can transmit to the next generation and therefore the better hunting technique is inherited.

Model five, based on state variable E, in Figure 3.8, combines the steady increase of new possibilities by new sexual recombinations and mutations. The sixth model, F, combines learning and steady increase of new possibilities by new sexual recombinations and the learning. The seventh model, G, is based on a combination of mutation with learning, and finally the eighth model, H, considers all three mechanisms. The equations in STELLA formulations for the models in Figure 3.7 are shown in Table 3.1 and for the models in Figure 3.8 in Table 3.2. It is not possible to quantify these mechanisms or, expressed differently, the models cannot be calibrated and validated because we do not have any observations of the quantitative role of these mechanisms. Therefore, for B, C and D it was chosen to obtain approximately the same values after 1200 generations—number of generations is the time unit—because that would facilitate the interpretation

Table 3.1 The equations for the model A, B, C and D in Figure 3.7

$A(t) = A(t-\mathrm{d}t) + (\text{increase1} - \text{decrease1}) * \mathrm{d}t$
INIT $A = 10$

INFLOWS:
increase1 = 1
OUTFLOWS:
decrease1 = $0.1*A$
$B(t) = B(t-\mathrm{d}t) + (\text{increase2} - \text{decrease2}) * \mathrm{d}t$
INIT $B = 10$

INFLOWS:
increase2 = 1 + 0.000015 * TIME
OUTFLOWS:
decrease2 = $0.1 * 10 + 0.02 * (B - 10)$
$C(t) = C(t-\mathrm{d}t) + (\text{increase3} - \text{decrease3}) * \mathrm{d}t$
INIT $C = 10$

INFLOWS:
increase3 = (1 + RANDOM(0.000, +0.025))
OUTFLOWS:
decrease3 = $0.1^* 10 + 0.02^* (C - 10)$
$D(t) = D(t-\mathrm{d}t) + (\text{increase4} + \text{Learning} - \text{decrease4}) * \mathrm{d}t$
INIT $D = 10$

INFLOWS:
increase4 = 1
Learning = $0.012 * (D - 9.5)$
OUTFLOWS:
decrease4 = $0.1 * 10 + 0.02^* (D - 10)$

Table 3.2 The equations for the models E to H in Figure 3.8

$E(t) = E(t - dt) + (\text{increase5} - \text{decrease5}) * dt$
INIT $E = 10$

INFLOWS:
$\text{increase5} = 1 + 0.000015 * \text{TIME} + \text{RANDOM}(0.000, +0.025)$

OUTFLOWS:
$\text{decrease5} = 0.1 * 10 + 0.02^{*}(E - 10)$
$F(t) = F(t - dt) + (\text{increase6} + \text{Learning6} - \text{decrease6}) * dt$
INIT $F = 10$

INFLOWS:
$\text{increase6} = 1 + 0.000015 * \text{TIME}$
$\text{Learning6} = 0.012 * (F - 9.5)$

OUTFLOWS:
$\text{decrease6} = 0.1 * 10 + 0.02 * (F - 10)$
$G(t) = G(t - dt) + (\text{increase7} + \text{Learning7} - \text{decrease7}) * dt$
INIT $G = 10$

INFLOWS:
$\text{increase7} = 1 + \text{RANDOM}(0.000, +0.025)$
$\text{Learning7} = 0.012^{*}(G - 9.5)$

OUTFLOWS:
$\text{decrease7} = 0.1^{*}10 + 0.02^{*}(G - 10)$
$H(t) = H(t - dt) + (\text{increase8} + \text{Learning8} - \text{decrease8}) * dt$
INIT $H = 10$

INFLOWS:
$\text{increase8} = 1 + 0.000015^{*}\text{TIME} + \text{RANDOM}(0.000, +0.025)$
$\text{Learning8} = 0.012 * (H - 9.5)$

OUTFLOWS:
$\text{decrease8} = 0.1 * 1 + 0.02 * (H - 10)$

of the results obtained by combinations of the three mechanism. The parameters in the four models of the combinations, E to H, are unchanged from B, C and D as it is mainly the intention for the models E to H to illustrate what effect it has to combine the mechanisms.

The results of the simulations are shown in Figures 3.9 and 3.10. Table 3.3 compares the different results with the idea to see if it would be possible to make some conclusions about the combinations of the mechanisms. Table 3.3 indicates which increase in eco-exergy is possible to gain by combinations of the mechanisms.

Provided that the models describe approximately the mechanisms, it is seen from the results in Table 3.3 that the sexual recombinations and the mutations are additive. Both mechanisms provide better solutions to survival under the prevailing conditions. It is therefore not surprising that they are additive. Whenever learning is included as a mechanism the result is more than additive. The learning as described requires that a surplus of time or resources is available to invest in learning, which explains why the learning has an effect that is more than additive when it is added to the effects of sexual

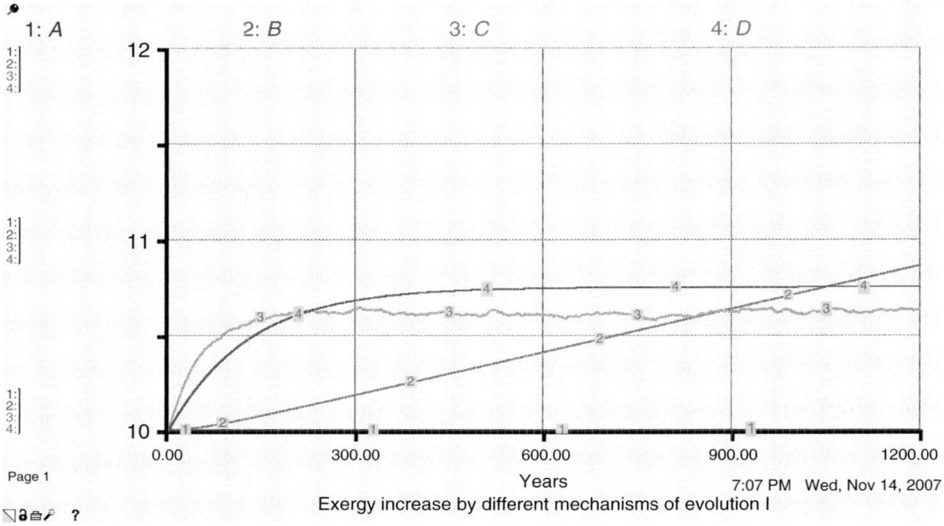

Figure 3.9 The results of models 1–4 with the state variables *A* to *D*.

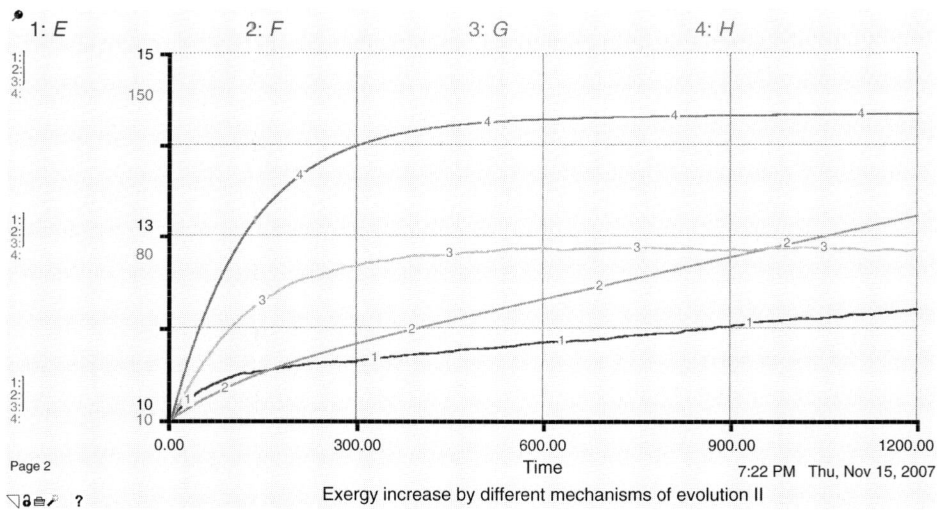

Figure 3.10 The results of models 5–8 with the state variables *E* to *H*.

recombinations and mutations. Heredity system 3 has therefore been increasingly significant as increasingly more complex organisms with a much better ability to learn have been able to benefit from this heredity system. Heredity system 4—the use of symbols—is probably working similarly but with an enhanced effect and would therefore probably also be more than additive with the results of sexual recombinations and

Table 3.3 Overview of the eight model results

Model #	State variable	Increase	Increase %	Comparison
1	*A*	0.00	0.0	
2	*B*	0.86	8.6	
3	*C*	0.64	6.4	
4	*D*	0.75	7.5	
5	*E*	1.50	15.0	$B+C=15\%$
6	*F*	2.77	27.7	$B+D=16.1\%$
7	*G*	2.31	23.1	$C+D=13.9\%$
8	*H*	116.8	1168	$B+C+D=22.5\%$

mutations. This explains probably the enormous rate at which the heredity systems 3 and 4 have evolved during the last hundred or thousand years compared with a period of the same length 10,000 or 100,000 years ago. The evolution today is completely dominated by the cultural evolution based on heredity systems 3 and 4, resulting in an evolution rate for heredity systems 3 and 4 that is very much higher than the evolution rate for heredity systems 1 and 2.

3.9 ILLUSTRATION OF THE ROLE OF CONJUGATIONS

The role of conjugation has been examined and illustrated by a software developed by Jørgensen, Chon and Recknagel, (in print). The software is based on a typical and interesting case study of conjugation. The software simultaneously illustrates how conjugation works and which advantages this process can offer to the evolution process; the details of the software is described below.

The model is developed to illustrate the dynamics of the gene–individual–population relationships to demonstrate survival of individuals with different types of genes under the constraints of food competition and toxic effects. The model elucidates three levels:

1. Gene regulation on physiological states of individuals such as metabolic efficiency and toxin susceptibility
2. Effects of feeding and toxins on individual's survival
3. Competition for food among individuals.

Individuals carry different gene information in determining metabolic efficiency and toxin susceptibility. While some individuals have an advantage in metabolic efficiency, other individuals have better resistance to toxins. Individuals carrying the genes move around on a lattice space to compete for the food with other individuals and with possible exposure to toxins. Health scores are generated as an attribute to each individual and to determine life processes (e.g., reproduction, death). For reproduction, binary fission was assumed, but exchange of genetic information could also occur through conjugation. Health scores are accumulated; positively with consumption of nutrients and negatively with exposure to toxins. The phenotypic outcome such as survival by competition would determine the gene levels of the population, while the gene levels control attributes of individuals. In the long run, the population has a higher chance of accumulating the most adaptable genes (i.e., survival of the fittest).

For simulation, an imaginary lattice with 800×800 units were assigned in two dimensions. The system was run in 1000 time steps and the software Visual Basic was used for constructing the IBM. The computer program is provided as individual-based model for the gene–individual–population relationships (IGIP). The model is available in Jørgensen, Chon and Recknagel (2008).

Variables

The attributes such as age, health score, and location of the individuals (x and y coordinates) were provided as attributes for individuals in the model. Additionally, health scores were used to address life processes of the individuals. Environmental factors, nutrients (food) and toxins were provided in the system. The nutrients in the lattice were accordingly consumed by the entering individuals and were resupplied regularly to the empty space. If the individuals were exposed to toxins while they move around, toxins were also consumed after exposure with the cost of decrease in the individual's health score. Toxins were also resupplied to the system on a regular basis to the empty space at the time when the nutrients were resupplied. At the population level, the total densities and densities of individuals in different types of gene information were used as variables.

Model structure and interaction

As stated above, life processes were presented in individuals, such as movement, consumption of nutrients, exposure to toxin, reproduction, and death. Individuals move according to random walk (one unit per time step in the Neumann type neighbours) if there is no constraint. Each interior site (i, j) (where $i = 2, \ldots, n-1$ and $j = 2, \ldots, n-1$) has eight immediate neighbour lattices $(i-1, j-1)$, $(i-1, j)$, $(i-1, j+1)$, $(i, j-1)$, $(i, j+1)$, $(i+1, j-1)$, $(i+1, j)$ and $(i+1, j+1)$. If a nutrient is located at one of the neighbour lattices, the individual moves to that lattice. If more than one nutrient exists in the neighbourhood area, coin tossing is carried out for determining the selection. Only one unit of nutrient is consumed by one individual at one time step. If the nutrients do not exist in the neighbourhood lattices, the individuals move at random. If toxins are located at the neighbour lattices, the individuals are exposed to only one of the toxin randomly.

The two types of genes for metabolic efficiency and toxin susceptibility were generated separately. The genes were assumed to control the degree of metabolic efficiency and toxin susceptibility at some fixed rates (e.g. 0.5, 0.25 and 0.1 for metabolic efficiency), while the maximum gene information was regarded as 0.5. Gene information was converted into phenotypic properties through health scores. The maximum score of metabolic efficiency and toxin susceptibility are assumed to be 20 points. Considering that the health score previously accumulated is 40 points and the individual has gene information for metabolic efficiency as 0.5, the updated health score will be 50 points (= 40 + (20 (maximum nutrient score) × 0.5). The health score would decrease similarly when the individual is exposed to toxins (20 (maximum toxin score) toxin susceptibility).

The health scores would be accordingly accumulated with history of food taking and toxin exposure as time progresses. If health score is >50 and age is older than 3 time steps, reproduction will occur, while individuals will die if the health score is ≤0 or age is more than 300 time steps. Individuals reproduce by fission, but conjugation could occur on a stochastic basis in case other individuals are located at the neighbour lattices. If two or more individuals are located at the neighbour sites, coin tossing is carried out for determination of the counterpart of the conjugation. Only one type of genes is selected for exchange at random. Suppose that an individual (say *A*) has 0.5 for metabolic efficiency and 0.5 for toxin susceptibility as gene information and *A* conjugates with another individual (say *C*) with 0.1 metabolic efficiency and 0.05 toxin susceptibility. If the gene for metabolic efficiency is selected for conjugation after coin tossing, the individual *A* will have 0.1 for metabolic efficiency and 0.5 for toxin susceptibility, while individual *C* will carry 0.5 for metabolic efficiency and 0.05 for toxin susceptibility. For simplicity of modelling, we assumed that conjugation occurs at 100% when the individual at the centre lattice was placed closed to the neighbouring individuals.

Parameters and input

The variables presenting individual (e.g., health score, location, age) and population properties (e.g., density in different types) were updated and provided to the system in each time step. In addition, nutrients and toxins were provided as environment factors in the system. Only one, either nutrient or toxin, is allowed per lattice. The parameters for supply of nutrients (initially 20% of the total lattice and resupply in 10% of the empty space in each 100 time step) and toxins (initially 20% of the total lattice and resupply in extra 1% of the empty space in each 100 time steps) are also incorporated to the system. For operation of the model, initial conditions for input variables were provided: Age (0 time step, (range: 0–300)) and health score (40 points, (range: 0–100 points)). Three types of individuals (10 in each type) with different gene information were assigned as initial population, for instance as listed below:

Type *A*: metabolic efficiency—0.5; toxin susceptibility—0.5
Type *B*: metabolic efficiency—0.25; toxin susceptibility—0.25
Type *C*: metabolic efficiency—0.1; toxin susceptibility—0.05.

Type *A* is strong in metabolic efficiency (0.5) and weak in toxin resistance (high in toxin susceptibility (0.5)), while Type *C* is the opposite of Type *A*. Type *B* is in the intermediate range in metabolic efficiency (0.25) and toxin susceptibility (0.25). We provided a small variation in Type *C* with a lower value of 0.05 for toxin susceptibility compared with 0.1 for metabolic efficiency. Each individual was placed at random in the 2D lattice space initially.

Output and model results

The model results are a summary of the more detailed results presented in Jørgensen, Chon and Cho (submitted). The time development of population size with different types of gene information was calculated as output. Through conjugation, nine types were produced in combination of the two types of genes:

A-A: metabolic efficiency—0.5, toxin susceptibility—0.5
A-B: metabolic efficiency—0.5, toxin susceptibility—0.25
A-C: metabolic efficiency—0.5, toxin susceptibility—0.05
B-A: metabolic efficiency—0.25, toxin susceptibility—0.5
B-B: metabolic efficiency—0.25, toxin susceptibility—0.25
B-C: metabolic efficiency—0.25, toxin susceptibility—0.05
C-A: metabolic efficiency—0.1, toxin susceptibility—0.5
C-B: metabolic efficiency—0.1, toxin susceptibility—0.25
C-C: metabolic efficiency—0.1, toxin susceptibility–0.05.

Figure 3.11 shows the population dynamics obtained by the simulations. Dominance types selectively appeared as the time progressed. Without conjugation (3.11a), no recombination of genes occurred, and the three original types (*A-A*, *B-B* and *C-C*) were only produced. In this case, Type *A-A* that carries genes for maximum rate of metabolism (0.5) and the minimum rate of toxin resistance (i.e. the highest level of toxin susceptibility (0.5)) was dominant first for more than 300 time steps compared with the other types (Figure 3.11a). At some lower levels, Type *B-B* was established, too. The other Type (*C-C*), however, was not established by the simulation. The simulation results revealed that the system constrained by food competition and toxin exposure could serve as a selection force for determining dominance of the traits in ecotoxicology.

The organisms with higher efficiency in food consumption appeared to be advantageous since nutrients were supplied at a relatively higher level initially. Densities of the *A-A* and *B-B* types, however, did not persist for a long time. They temporarily dominated and declined soon. In the middle of simulation, the total density remained at low levels for a substantial period. This is due to the depletion of food at this stage. Consequently, population could not recover readily after the nutrients were exploited by the organisms. However, populations started to recover again later around 600 time steps. Extra nutrients were accumulated during the periods of low densities after the dominant type had been declined in the early period. The same types, Type *A-A* and Type *B-B*, became dominant again, ensuring that Type *C-C* was not adapted to the system as well as the two other types. But both types were not recovered strongly, and started to decline again around 1000 time steps. The gene of Type *C-C* is clearly not competitive. The genotype–phenotype relationships are, however, complicated and further study is required regarding the gene–environment causality relationships.

When conjugation (100%) was allowed for gene exchange, however, population dynamics appeared differently (Figure 3.11b). Initially, Type *A-A* was dominant similar to the case without conjugation, but Type *A-C* gradually became dominant to exceed Type *A-A* eventually. Type *A-C* was obtained from conjugation and continuously

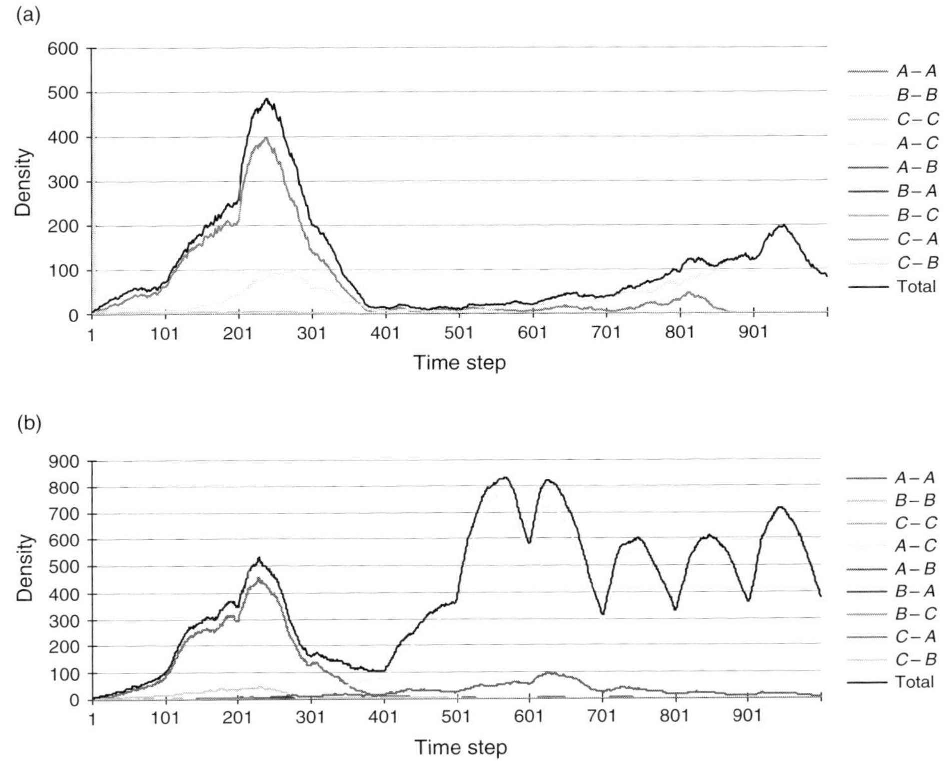

Figure 3.11 Results of the described test simulations: (a) without conjugation, (b) with 100% conjugation. The evolutionary advantages of conjugations are seen very clearly from the results. After 1000 time steps the total density representing biomass and eco-exergy is about three times higher with conjugation than without conjugation. More detailed results are shown to illustrate the role of conjugation.

increased as the time progressed. This is understandable: Type *A-C* carries the most adaptive gene information (maximum both for metabolic efficiency and toxin resistance). At the later period of simulation, Type *A-C* was completely dominant, prevailing over the entire lattice space. The population also showed periodic changes. This maybe due to coupling with the periodic nutrient resupply as stated before.

The model results show that IBMs are favourably applied when addressing complex phenomena in ecotoxicology. Comprehensive understanding was available on addressing the properties residing in the complex gene–individual–population relationships. Interaction with other individuals (i.e., competition) and environmental factors was demonstrated in IBMs, being efficiently fabricated with the model structure interlinking the upper and lower levels in the hierarchy of biological systems. The phenotypic traits were sensitively dynamic to the constraints of the genotypes. However, this study only proposed an initial stepping stone in IBMs towards the gene–individual–population

Table 3.4 Model results by conjugation 0, 25, 50, 75 and 100%

Conjugation (%)	Species richness	Total density	Shannon's index
0	1	402	0
25	3	417	0.985
50	3	412	1.083
75	3	411	1.016
100	3	418	1.013

relationships in ecotoxicology. Further study is required in the future on many different aspects regarding computational analysis on system dynamics residing in the individual–population relationships, quantitative characterization of the emergent properties produced from the model, biological/ecological experiments linking genes-behaviour-population relationships, and technical development in IBM construction.

The model was used to carry out a more comprehensive series of simulations that could illustrate in more detail the role of conjugation. Table 3.4 shows the model results based on the metabolic efficiency and the toxin susceptibility of *A-A*, *B-B* and *C-C* and with different percentage of conjugations. The results are obtained by averaging the results from the last five iterations with 100 iteration time intervals upto time = 7000, i.e. the average results after time 6600, 6700 and 7000.

The results are consistent with the results in Figure 3.11, in the sense that 100% conjugation yields higher density and therefore also biomass and eco-exergy. The species richness after time = 1000 was higher by 100% conjugation, namely 9. This discrepancy from the results in Table 3.4 is probably due to the longer simulation time applied for the results in Table 3.4. Longer time will inevitably mean a higher probability for outcompetition of more species. The Shannon index, species richness and the density are not increased, completely parallel or linearly to an increase in the percentage of conjugation, although the conjugation clearly in all cases gives higher density and higher species richness and Shannon index. The model is relatively complex and two important properties are interacting simultaneously, namely metabolic efficiency and toxin susceptibility. At the same time, there is an element of randomness and the initial conditions and the spatial distribution play also a role, which may explain the obtained results. The main conclusion is clear: introduction of conjugations implies a higher eco-exergy and a higher biodiversity. Conjugation gives, therefore, the evolution a significant push forward towards higher eco-exergy and more possibilities for moving further away from thermodynamic equilibrium. A higher species richness implies more possibilities for further genetic improvements by mutations and sexual recombinations and more possibilities for establishment of larger ecological networks. Similarly, sexual recombination affords a vast scope for change and for new evolutionary possibilities, yet the genetic integrity is still maintained. The diversity that, for instance, 1000 individual of a mammal species with 25,000 genes and totally 6 billion bits of information is astronomic. The introduction of conjugation and later sexual recombination in the evolution have given an enormous rise in the variability that has been of utmost significance for all the evolutionary processes.

4

Extreme environments

4.1 INTRODUCTION

Time after time, we have been surprised by the wide spectrum of extreme environments that are carrying life on the Earth. How can life exist several kilometres under the surface of the sea at high concentrations of sulphide and very hot water? The organisms have no light and are exposed to an enormous pressure. How can life survive in tidal ecosystems with the changing conditions from no water to complete coverage of water? How can life exist in deep caves with complete darkness? The reason could be that life will always be present where there are conditions for life, even if the conditions are very harsh. If life has sufficient time to develop it will inevitably develop even under harsh or extreme conditions, provided that some basic requirements or needs are fulfilled (see also Sections 2.5 and 5.2). If it is the case, we should expect at least simple one cellular life in our solar system, maybe on Mars, maybe on Europa, one of Jupiter's moons and maybe on one or two of Saturn's moons (see the discussion in Section 5.2). We cannot, however, exclude that the possibilities to have life under extreme conditions are determined by the rich variety of organisms that we can select from the Earth. Due to the high diversity of species on the Earth, there is a high probability that one or more of the species can modify its properties to deal with the conditions in an extreme environment, particularly if sufficient time is available. These factors are illustrated in Figure 4.1. In other words, the reason why we have life in this extreme environment on Earth could be the richness of the many different life forms that have already been established. It is also in accordance with the exponential increase of the diversity, which we shall discuss in more detail in the Third Movement.

The first life on the Earth was probably formed in very gentle environments—or maybe coming from the outer space, and later it has been possible with the enormous diversity in species to find solutions to how life could evolve under extreme conditions.

We cannot, however, conclude which of the following two possibilities would give the most correct explanation for life in extreme environments:

1. Life will inevitably emerge everywhere when the conditions are suitable for life, both under harsh or gentle conditions.
2. Life under harsh conditions is only possible if life has had sufficient time to form many different life forms among which some can be modified to meet harsh life conditions.

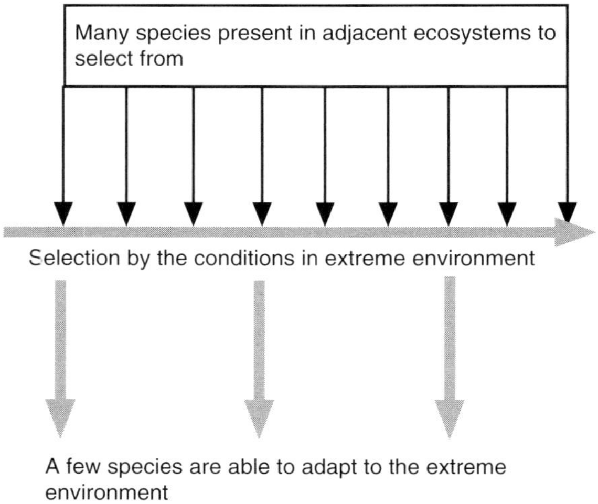

Figure 4.1 Life under harsh conditions is only possible if life has had sufficient time to form and have many different life forms in adjacent ecosystems among which some can be modified to meet harsh life conditions.

We are, probably, mostly inclined to accept the explanation in Figure 4.1 in the light of exponential growth of the diversity. Later, if we find life in our solar system under harsh conditions, we are, however, closer to give a proper explanation for life that we can find on the Earth by the most extreme environment. It cannot also be excluded that one has to use a combination of both explanations.

Let us, however, give a description of life forms in some of the most extreme environment, because it would give us a better understanding about which factors are important for the evolution of life.

4.2 THE DEEP SEA

Until recently, deep sea was considered to be an ecosystem with a low abundance of organisms due to the extreme limitation of food supply. There is, however, one habitat in the deep sea where the density of life is almost equal to what is found in many other marine ecosystems. It is the system of hydrothermal vents or deep-sea hot springs. The vents are found along the ridges at the bottom of the ocean where Earth's crystal parts are spreading apart (Childress et al., 1987). They consist of densely packed animal life several kilometres below the surface. They are living in total darkness; however one can find giant tube worms, for instance *Riftia pachyptila*, as long as 1 m, large white clams 30 cm in length (*Calyptogena magnifica*) and clusters of mussels, mainly *Bathymodiolus thermodphilus*, forming thick aggregations around the hydrothermal vents. Also shrimps, crabs and fish can be found. Photosynthesis is impossible at the depth of the vents. What is the source of energy, because an energy source is absolutely necessary for all ecosystems? Can the density of life be explained by the temperature? The temperature of vent water is 10–20 °C in contrast to most deep sea water which is very cold, namely 2–4 °C. The warm temperature does not, however, explain the unique life of the vent ecosystems. The explanation is that these underwater springs are rich in hydrogen sulphide as many springs on land. Some of the vents are called black smokers, because they are completely black in colour owing to the presence of metal sulphide. The sulphide-rich habitats support a large number of bacteria. They are autotrophs, and do not use Sun as energy source, but derive the energy by oxidation of hydrogen sulphide. The sulphur bacteria have substituted the green plants closer to the surface. Whereas green plants are photoautrophs, the sulphur-oxidizing bacteria are chemotrophs, i.e. capable of using inorganic energy source to drive carbon dioxide fixation. Figure 4.2 shows a typical and possible food chain close to a deep sea vent (Colaco et al., 2007).

The hydrothermal vent communities consist of unusual invertebrate species such as a dense cluster of tube worms, vent mussels and vent crabs. The tube worm is essentially a closed sac without a mouth and digestive system. At its anterior tip there is a red gill like plume where oxygen, carbon dioxide and hydrogen sulphide are exchanged with the ambient sea water. The animals consist mainly of a thin-walled sac that contains the internal organs. The largest of them is the trophosome, which occupies most of the body cavity. It contributes significantly to the worm's nutrition and is colonized by a vast number of the sulphur-oxidizing bacteria. The tube worm and the bacteria have established an endosymbiotic relation. The tube worm receives reduced carbon molecules

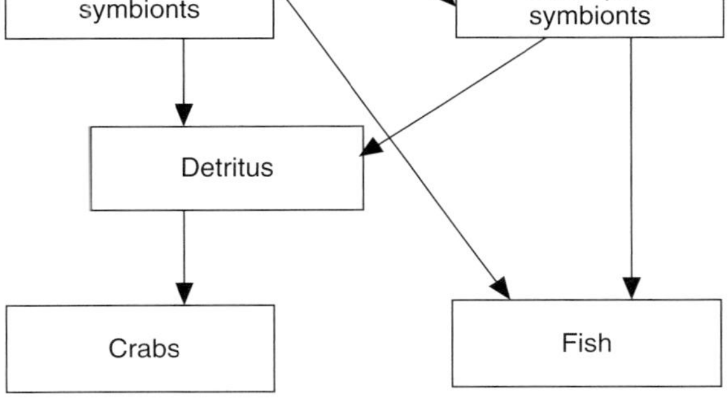

Figure 4.2 The food chain in the deep sea at hydrothermal vents. The energy source is hydrogen sulphide, oxidized by bacteria. The tube worms utilize the chemical energy of hydrogen sulphide by use of endosymbiotic bacteria.

from the bacteria and in turn provides the bacteria with the raw materials needed to fuel its metabolism: carbon dioxide, oxygen and hydrogen sulphide (Colaco et al., 2002).

The crucial question is how the fauna of the vent can survive the high concentration of the toxic hydrogen sulphide that is able to block the respiration? Studies of the tube worm have revealed that sulphide has no effect on oxygen binding and that the worm's respiration is substantial even in the presence of sulphide concentrations that are lethal to most animals. The tube worms can extract sulphide from vent water and they have a blood component, the worm's high molecular weight haemoglobin, which is able to bind hydrogen sulphide stronger than the sulphide-sensitive cytochrome *c*oxidase (Stewart et al., 2005). The haemoglobin can bind oxygen and hydrogen sulphide simultaneously but at different sites. The large white clam and the mussels have also developed a symbiotic relation to the sulphur bacteria. The bacteria are not, however, in the internal organ but in the gills, where they can readily obtain oxygen and carbon dioxide from the respiratory water flow. The metabolic plans are however the same, namely that the bacteria oxidize sulphide and supply the clam with fixed carbon compounds. The clams concentrate the sulphide in their blood. The level is orders of magnitude higher than in the concentrations in the ambient water. Apparently, the clams absorb sulphide through their large elongated feet, which extend into the hydrothermal vents, where the concentrations of sulphide are highest. After absorption through the clam's feet, sulphide is transported to the bacteria in the gills by the blood. The transport is carried out by a special high molecular weight protein, which is able to protect the sulphide against oxidation on the way to the gills and also protects the haemoglobin and cytochrome *c*oxidase. The binding of sulphide to the protein is reversible. It is off-loaded to the bacteria in the gills and oxidized to provide energy.

The tube worm, the clam and the mussels owe their ecological success to their symbiosis with the sulphur bacteria (Calaco et al., 2007). Many of the smaller vent animals lack symbionts. They obtain their nutrients either by filtering particular food such as bacteria from the water or by feeding on animals that contain symbionts. Vent crabs, for instance, feed on the respiratory plume of tube worms.

Many of the animals in the deep sea can produce light by chemical reactions of special proteins, for instance fish, octopus and squid. There are also light-producing bacteria. It gives clearly a great advantage to produce light, where there otherwise is complete darkness.

The hydrothermic activity in the deep sea is not only the energy source directly or indirectly for many life forms, but it is also an important source of minerals for the entire marine life.

4.3 INTER-TIDAL FISHES

The inter-tidal zone is twice a day cutoff from the open ocean. The water that remains is left in isolated tide pools or under rocks, else it forms mud flats. When the water returns at high tide, the inter-tidal zone is submerged to rejoin the ocean. Any animal that lives in the inter-tidal zone must be able to spend much of its time either completely out of water or at least partly exposed to air. The periods during which it can feed are set by the tidal cycle, and it implies that it must withstand wide variations in water chemistry and a constant wave action and turbulence, characteristic for the zone. Barnacles, limpets, periwinkles and fish are found in these very variable ecosystems. At low tide, many species live under rocks or within clumps of vegetation. Many of them are camouflaged by protective colouration. Species that live within the algal vegetation usually have colours that match the surrounding seaweed, for instance the gunnels *Apodischthys flavidus* and *Xererpes fucorum*.

Inter-tidal fish are rarely longer than 30 cm. The small size enables them to occupy holes, crevices and spaces under rocks (Horn and Gibson, 1988). It reduces the risk of being swept away by surge or waves. The skin is tough and can withstand repeated scraping against the rocks. Many of the fishes secrete large amounts of mucus, which may provide lubrication. Many of the inter-tidal fishes have negative buoyancy, which enables them to lie effortlessly on the bottom, where cover is nearby and water velocities are lowest. Many inter-tidal fishes can tolerate considerable loss of water—as high as 60% of their total water content, for instance *Pherallodiscus funebris*. This is a higher loss of water than that can be tolerated by most of the amphibians. Certain animals, for instance the tropical mudskippers, are virtually amphibians. The rate of desiccation in some species are slowed down by anatomical features as a thickened epidermis and the presence of mucus-secreting cells in the skin, but inter-tidal fishes apparently have no physiological mechanism for reducing or preventing water loss. It seems, however, that behaviour plays an important role in survival, for instance by returning frequently to the water to avoid excessive drying of the skin and respiration surfaces.

All inter-tidal fishes must obtain oxygen even when they are out of water. The problem is not oxygen, because air contains of course oxygen, but the problem is that the gills tend to collapse in the air. This problem is solved by shorter and thicker gill filaments, preventing the collapse of the gills enabling them even to breathe air. Another problem is the strong diurnal fluctuations in the amounts of dissolved oxygen and carbon dioxide in the pool water. The seaweed photosynthesis can supply amounts of oxygen exceeding the respiratory demand resulting in extreme high oxygen concentrations and low carbon dioxide concentrations. At night, when there is no photosynthesis, the oxygen may be depleted by the pool's plants and animals and the level of carbon dioxide is high while the oxygen concentration is very low. Some tide-pool fishes can regulate the oxygen consumption and can thereby compensate for these fluctuations in the concentrations of oxygen and carbon dioxide.

It is possible to conclude that the fishes that live between the tides and from time to time isolated in pools and on mud flats have been able to develop anatomy, physiology and behaviour that fit to the rigorous habitat.

4.4 CARNIVOROUS PLANTS

The luring, capturing and digesting mechanisms that have evolved in some plants enable them to capture insects to augment their supply of nutrients and thereby survive in habitats where few other plants can live. The carnivorous plants can be divided into two groups according to their methods of catching prey. They are either active or passive trappers (Heslop-Harrison, 1978). The active trappers catch hopping or crawling insects. When the prey touches the leaf, tactile hairs are agitated and a closing mechanism is triggered.

For the Venus's flytrap, the two sides of the leaf quickly move together closing the trap. For the bladderwort, the flap of the tissue that forms the door swings open and the bladder expands suddenly to draw in both water and prey.

The passive trappers, for instance the pitcher plant, lure the prey to a slippery edge and the prey falls into a pool of digestive fluid and cannot climb out. Another passive trapper, the sundew, has attractively coloured leaves. When small flying insects touch the secretion globules on the leaf surface, digestive enzymes are secreted. The nutrients derived from the captured prey enter the leaves in a surprisingly faster rate.

It has been shown by numerous experiments that the carnivorous plants performed better when provided with prey. The main function of the carnivorous habitat is to provide scare nutrients. It is therefore expected that they inhabit environments where such supplementation would be beneficial, which is also in accordance with the observations. The carnivorous plants are encountered most frequently in nutrient-poor ecosystems, heats, bogs, impoverished soil in forest openings and on clay soil associated with weathered limestone.

The carnivorous plants occupy extremely narrow ecological niches. The range of adaptations found in the carnivorous plants is furthermore completely according to Darwin's absorption with carnivorous plants as examples of evolutionary virtuosity. Four hundred of the about 300,000 species of flowering plants are known to be carnivorous. They belong to 13 or so of genera of 6 families.

The supplemental nutrients available to the carnivorous plants from the digestion of the preys offer them special advantages, particularly of course in environments where nutrients are scarce. Both nitrogen and phosphorus seem to play an important role in this context. But are these advantages counterbalancing the costs? The plants have to invest energy in the synthesis of digestive enzymes and other secretion products and in the plant's elaborate structural adaptation. The carnivorous plants are, however, found in places with adequate carbon sources, access to water and where there are no limitations of the photosynthetic energetic resource. Therefore, the question is not about the energy cost but about survival in places where no non-carnivorous competitors can intrude. Whatever the energy cost may be, the investment is justified by the survival possibilities.

4.5 WINTER MOTHS

There are moths that can fly, feed and mate at near-freezing temperatures. How is that possible? Winter means commonly death to adult insects. Most owlet moths are active only on warm summer nights. They die off as the winter approaches leaving behind eggs and caterpillars or pupae that remain inactive until the next spring. The winter noctuids, for instance *Eupsila* and *Lithophane*, in contrast, emerge as adults in the fall or late winter, when they feed, mate and lay their eggs before dying in the spring. Their caterpillars feed on the early spring eating buds of forest trees and are then quiescent throughout the summer. How do the winter moths survive when other moths die? What enables them to avoid freezing and make it possible for them to fly and see food in the cold? The winter moths are endothermic. To fly they must keep their wing muscles heated at about 30 °C. They have the same temperature as the air when they rest, while they heat themselves endothermically before flying, even if the air temperature is close to the freezing point. Furthermore, the winter moths are able to shiver. The heating requires a considerable amount of energy. Their metabolic rate is however not high. The heat energy required in winter moths is generated by shivering for more than half an hour to reach a thoracic temperature of 30 °C. During flight, the moths repeatedly stop, shiver and thereby warm up again. Owing to lack of low-cost energy sources, the moths seem to be selective about the conditions under which they shiver. The animals do not heat themselves to keep warm. If they do not need to fly they neither warm up nor resist cooling after a flight. The lower the air temperature, it is less likely that the moths shiver (Heinrich, 1987). While most of the moths shiver at 17 °C, only half in a study shivered between 5 and 8 °C. This apparent emphasis on energy conservation makes sense when it is considered that at an air temperature of 0 °C its shivering would exhaust the contents of its stomach – about 4 mg of sugar – in just 35.2 min.

Because the moths do not have highly specialized mechanisms for producing extra warmth, they must have an effective way of retaining heat. The moths are well insulated by a coat of dense pile, which is a derivative of scales. They are also able to prevent heat from diffusing out of the thorax into colder areas of the body. All endothermic insects such as dragonflies, bumblebees and honey bees have such an ability. The animals retard heat flows to the head and the abdomen and eliminate leakage to other extremities such as legs and wings.

The remarkable ability of winter moths to be active at low temperature illustrates how evolutionary alterations in anatomy, physiology and behaviour can add up to success in a new environment. The winter moths are still quite similar to their close relatives, but the small differences adapt the insects to winter living.

4.6 DEEP CAVES

In deep caves, where no light can penetrate, live millipedes, spinners, scorpions, worms and amphibians without eyes. The can move, kill, mate and feed in complete darkness often in a poisonous atmosphere. About 7700 species have been found up to now, but it is expected that the number will increase significantly during the coming years. How can they survive under these extreme circumstances?

The species found in the deep caves have been adapted to the extreme environment in the deep cave. A moss scorpion uses a claw instead of the tail to inject its poison, but it is a formidable predator. A 30-cm-long scolopendra is able to kill and eat bats. The spinner, *Titiotus*, is twice as big as its relatives on the surface and does not use nets but is able to capture the prey with its legs.

The cave species have some similarity with their surface relatives, while the differences can be considered as adaptations to the special conditions in the deep caves. It is therefore most probable that the species in the caves have not been evolved in vacuum, but are evolved from the surface species. The life in the deep cave has therefore the surface species as the prerequisite, and would not have been able to habit the deep caves without the surface species been evolved first to the extreme environment in the deep caves.

4.7 SUB-GLACIAL LAKES

There are about 70 sub-glacial lakes in Antarctica. The largest is called Vostok Lake. It is 14,000 km^2 and 400 m deep. The lake is 4000 m under the surface, where temperatures of about −85 to 90 °C have been measured. The lake has probably been isolated from the Earth in more than 35 million years. The water is probably liquid water and not ice due to the high pressure. A Russian team of researchers have planned to examine what life forms, if any, the lake is able to show. The temperature and the isolation by a thick layer of ice from the rest of the planet are characteristics similar to what has been found on Jupiter's moon, Europa. If the lake has life, it is probably unknown life forms and it will probably be possible to conclude that life will occur sooner or later, wherever the conditions are able to bear feasible life forms. If the Vostok Lake has life, there is a high probability that Europa has life, and search for life on Europa will be intensified. The results of the Vostok Lake investigations will probably be published in late 2009 or early 2010. Unfortunately, several researchers are concerned about the possibilities to exclude contamination as a result of the investigation methods by microscopic life forms on the Earth. The results will be very interesting and create discussion about the possibilities to find life in our solar system. See also the web sites www.newzeal.com/thme/bases/Russia/Vostok and www.bbc.co.uk/scinece/horison/2000/vostok/stml.

Second Movement

History of the Biological Evolution from a Thermodynamic Point of View

Eco-exergy as a Criterion of Selection in the Macroevolution

5

The evolution of the universe

5.1 THE COSMIC EVOLUTION

Shortly (10^{-43} s) after the big-bang the temperature was 10^{32} °C. The universe consisted of quarks and gluons 10^{-11} s after the big-bang. The quarks and gluons became locked up inside the protons and neutrons 10 µs after the big-bang. The temperature has decreased to 2 trillion °C. The elementary particles, electrons, protons and neutrons, formed about 380,000 years after the big-bang, the first being the two light elements hydrogen (75%) and helium (25%). The formation of galaxies in the form of interstellar clouds started less than 1 billion years after the big-bang. The stars were formed about 10^{10} years ago and the planets 2–4 billion years later (Hawking, 2001).

The universe is steadily expanding according to Hubble's Law (Hawking, 2001). In accordance to the mysterious black energy the rate of expansion is increasing. The expansion of the universe implies that the distances between galaxies are increasing, which will create more order due to a mass distribution with a smaller probability. Kullback's measure of information will therefore increase (see Equation 1.20, Chapter 1). The corresponding increase in exergy is found to be in the order of 10^{57} J/yr (Jørgensen et al., 1998):

$$\Delta Ex = NkT[\ln aR/n_g r_g - \ln R/n_g r_g] = NkT \ln a \quad (5.1)$$

where N is the number of particles in the universe, 10^{81}, R is the radius of the universe, a represents the expansion factor $R \rightarrow aR$ (1/yr), k is Boltzmann's constant, 1.4×10^{-23}, n_g is the number of galaxies and r_g the average radius of the galaxies. T is the absolute temperature of the nucleons (10^9 K). a represents the expansion factor for one year $a = 1 + 1/H$, where H is Hubble constant $= (14 \times 10^9 \text{ years})^{-1} = 7 \times 10^{-11}$ 1/yr. As $\ln(1 + 7 \times 10^{-11})$ with good approximation is 7×10^{-11}, the exergy gained by expansion is found as

$$10^{81} \times 1.4 \times 10^{-23} \times 10^9 \times 7 \times 10^{-11} = 10^{57} \text{J/yr} \quad (5.2)$$

This calculations do not consider the dark matter, which outweighs normal matter by a factor of 6 (Caldwell, 2006). The exergy gained by expansion is therefore estimated to be 7×10^{57} J/yr. Furthermore, it is (see Lloyd, 2007) estimated today that the universe has expanded to 42×10^9 light years not only to the 14×10^9 light years, that correspond to the age of the universe. The expansion has therefore been at a rate three times the speed of light. If it is the case, should we furthermore multiply by approximately three to get the total increase in exergy due to the expansion. It means that the gain in exergy is about 2×10^{58} J/yr.

It is estimated that about 2×10^{12} stars are formed every year (Greene, 2000). It is found by dividing the total number of stars, about 2×10^{22}, with their life time—about 10^{10} years. If it is estimated that a star has in average the mass of the Sun, 2×10^{33} g, and that a star in its formation phase contains mainly hydrogen, it is 2×10^{33} moles. A typical density of a star is 1.5 kg/l or 1500 kg/m^3, while the average density of an interstellar cloud is 10^{-19} kg/m^3 (Hawking, 2001). The exergy due to the difference in density can be found by use of Equation 1.17, Chapter 1, presuming a temperature of 10^7, corresponding to the temperature needed for hydrogen fusion processes to be realized:

$$RT\,2 \times 10^{33} \ln 1500/10^{-19} = 8.3 \times 10^7 \times 2 \times 10^{33} \times 320 = 8.5 \times 10^{42}\text{J} \qquad (5.3)$$

In addition, the high temperature has also a ‘‘working capacity’’ corresponding to the usually applied equation: gradient $\times$ extensive variable $=$ temperature difference $\times$ heat capacity. Presuming that the interstellar temperature is 15 K and the heat capacity of hydrogen as a monatomic gas is $3R/2$ (see any textbook in physical chemistry), the work capacity can be found as

$$(3 \times 8.3/2)2 \times 10^{33} \ln 10^7/15 = 3.33 \times 10^{35} \qquad (5.4)$$

which is negligible compared with the contribution found in (5.3).

The increase in exergy due to formation of stars is therefore about 8.5×10^{42} J times 2×10^{12} stars/yr $=$ about 1.7×10^{55} or almost 1000 times less than the exergy gained by the expansion.

The interpretation of the second law of thermodynamics that the universe evolves towards the ‘‘heat death’’ and decreased order should therefore be modified. Layzer (1988) uses the expression ‘‘potential information’’, I. We have that

$$I + S = S_{\text{eq}} = S_{\text{max}} \qquad (5.5)$$

S_{eq} is the entropy at thermodynamic equilibrium corresponding to the heat death. It is therefore also the maximum possible entropy of the universe, denoted S_{max}.

If we express the same idea by exergy, we have that

$$\text{Ex}_{\text{universe}} + \text{exergy}_{\text{expansion}} = \text{exergy}_{\text{available}} \qquad (5.6)$$

where $\text{Ex}_{\text{universe}}$ is the exergy stored today in the stars and planets (included in the living organisms, although it is most probably minor), $\text{exergy}_{\text{expansion}}$ is the exergy gained by the expansion and $\text{exergy}_{\text{available}}$ is the exergy that the universe has available to do work. $\text{Ex}_{\text{universe}}$ is of course decreasing according to the radiation produced by the more than 10^{22} stars + other exergy-consuming processes. The Milky Way Galaxy has 2×10^{11}

stars, which are responsible for most of the exergy flow. On an average, a star has a yearly loss of exergy similar to the Sun corresponding to about 1.5×10^{37} J/yr. Supernovas, black holes and so on are considered when we use this loss of exergy per star. It means that for the Milky Way Galaxy the loss of exergy is 3×10^{48} J/yr. With about 10^{11} galaxies, the decrease of exergy stored in the stars is therefore close to the exergy gained by the expansion, namely about 3×10^{59} J/yr $> 2 \times 10^{58}$ J/yr.

The "heat death" moves therefore probably not closer or at least at a very low speed, although the universe produces an enormous amount of entropy per year and uses an enormous amount of exergy due to radiation. It is the general perception that the rate of universe expansion is increasing due to what is called dark energy. In the coming years, astrophysics will probably be able to give a better explanation than just to apply the term dark energy for something unexplainable. Anyhow, if—what is most probable according to the last hypotheses in astrophysics—the universe is expanding faster and faster, it will imply that the $exergy_{available}$, the total exergy of the universe may even increase. The pessimistic prediction about the "heat death" is therefore wrong.

It is interesting in this context that the exergy flow density in the sequence, galaxy → stars → planets—and later in this movement, it will be shown valid also for the biological evolution towards more and more complex life forms—increases, indicating a clear direction of development: towards increasing possibilities for a faster and faster development in the cosmic evolution due to a higher and higher exergy density flow, which seems in contradiction to the classical interpretation of the second law of thermodynamics of a development towards the heat death, that is a development towards a complete homogeneous universe without gradients and work capacity.

The galaxies were formed about 13 billion years ago, the first stars about 9–10 billion years ago and the first planets about 6–8 billion years ago. The solar radiation (which can be considered almost 100% exergy) amounts to 4×10^{26} J/s and its mass is 2×10^{30} kg giving an exergy flow density of 2×10^{-4} J/kg s, rooted in an enormous eco-exergy density in the order of 10^{13} J/kg. The Milky Way Galaxy has 2×10^{11} stars, which are responsible for most of the exergy flow, corresponding to about 10^{38} J/s, including supernovas, black holes and so on. The mass of the Milky Way Galaxy, inclusively the interstellar gas, dust and dark matter, is about 2×10^{42} kg. The exergy flow density is therefore 5×10^{-5} J/kg s. The solar radiation reaching the Earth is about 1.3×10^{17} J/s. The weight of the active part of the Earth—the atmosphere, the upper 100 m of the oceans and upper 20 m of the terrestrial ecosystems—is about 10^{19} kg; the exergy flow density for the Earth becomes 1.3×10^{-2} J/kg s. So, the eco-exergy flow density has increased about 260 times from the formation of galaxies to the formation of planets. Later in this movement, we shall look at other exergy flow densities and exergy densities to see if there is a general trend in the evolution towards higher exergy flow density and higher exergy density.

5.2 THE PROBABILITY TO FIND LIFE ELSEWHERE IN THE UNIVERSE

Is life on Earth unique? Or can we find life elsewhere in the universe? In accordance to the discussion in Chapter 1, life is a result (natural consequence) of an open system with sufficient input of energy/exergy, with a suitable composition of the elements that characterizes life (C, N, P, S, etc.), with a temperature of about 250–350 K, water and sufficient time. We will, according to these requirements, expect to find life on Mars or at least fossil of primitive life. We may also find life on the moon Europa, because volcanic activity ensures the right temperature in a certain depth and it has been assessed that water is present. The life we will find in our solar system besides the Earth will, however, inevitably be primitive life, as already pointed out, because the conditions on Mars and Europa have not been so favourable as on the Earth, at least not sufficient long time to allow the evolution from primitive life via multicellular life, invertebrates, vertebrates (fish and amphibians), dinosaurs, mammals and apes to *Homo sapiens* and latest to the technological man that masters the radio waves and can communicate wireless. But can we find human-like life with sufficient technology on other planets in the galaxy? If the answer is yes, it is a puzzle that we, after about 40 years listening to the space, still have not been able to identify radio signals from another planet in another solar system in the galaxy.

Drake has proposed the following equation:

$$N = R^* \times f_p \times n_e \times f_l \times f_i \times f_c \times L \tag{5.7}$$

where N is the number of other intelligent civilizations in our galaxy, R^* is the number of stars that are born per year (20 of which only half would have the right size—therefore 10), f_p is the fraction of stars with planet systems (Drake estimated that this factor is 0.5—theoretical calculations indicate 0.33; see Rasmussen, 2005), n_e is the fraction of planet systems that could bear life on one planet, which means that the temperature is right and that the surface is solid (opposite the gas planets Jupiter, Saturn, Uranus and Neptune) (0.15 according to Drake), f_l is the fraction of planets where life is possible and where life actually can be found (1.0) (when life is possible, it will always emerge), f_i is the fraction of planets where the conditions have been favourable sufficiently long time to evolve intelligent life (0.1—estimated by Drake), f_c is the fraction of planets with intelligent life that can communicate with radio waves (estimated to 0.5) and L is the time that a highly developed civilization is expected to survive. We can estimate L to 1000 years, 10,000 years or 100,000 years. If we select $L = 10{,}000$ years, we have that

$$N = 10 \times 0.33 \times 0.15 \times 1.0 \times 0.1 \times 0.5 \times 10{,}000 = 250$$

planets would have intelligent life that could communicate with us. If $L = 1000$ years, $N = 25$, and if $L = 100{,}000$ years, $N = 2500$. Why have we still not received any radio signal from these civilizations? Even with 25 planets with intelligent life would mean that we should have a high probability to have a technological civilization as a neighbour in a distance of about 5000 light years. Is it because intelligent life is self-destructive and that even $L = 1000$ years is too long? Should L rather be 200 years, indicating that we probably soon will destroy all life on the Earth by, for instance, lack of environmental concern or a nuclear global war?

Another explanation could be that any evolution of life inevitably meets several obstacles or evolutionary bifurcations, which only by chance can be bypassed. It may, for instance, be important that planets with long evolution time have a relatively heavy moon in the right distance to give tide water, which enhances the change of life forms between the water and the land, and stabilizes the inclination of the rotation axis—let us multiply with f_m. Among the three planets that have the right distance from the Sun—Venus, Earth and Mars—only the Earth has a heavy moon. Let us therefore tentatively estimate that $f_m = 0.33$. The Earth has experienced probably at least five major catastrophes that all have given the evolution a kick and a new chance to evolve towards intelligent life. If the Earth would not have had the catastrophe 65 million years ago where the mammals got a new chance, dinosaurs may still have been the kings of life on Earth. The most probable explanation for this catastrophe is that an asteroid has made an impact. An asteroid belt as in our solar system between Mars and Jupiter and many comets may be necessary to ensure a certain frequency of catastrophes in a solar system. There have also been catastrophes above 440 million years, 360 million years, 251 million years and 201 million years ago. The last two mentioned catastrophes were probably due to climatic changes, probably caused by volcanic eruptions. Let us introduce a factor f_a to account for the need for catastrophes to give the evolution new chances. Let us estimate that f_a is as high as 0.25, because there will probably be elsewhere in the universe many possibilities for major catastrophes.

Finally, the random climatic changes on Earth have been of importance for the evolution towards intelligent life. Let us mention two key illustrations of this important factor:

1. The right climatic changes and the simultaneous increase in the oxygen concentrations were prerequisites for the Cambrian Explosion (see also Section 8.2).
2. About 5 million years ago, the climate became drier and the rain forests in many parts of Africa changed to savanna. Without this change in climate, the anthropogenic apes would not have been forced to walk upright and *Homo sapiens* would not have been the result. It is also noticeable that the climatic changes in this case happened just when the anthropogenic apes were ready for a further evolution.

As previously mentioned, the evolution depends on the randomly changing life conditions. The changing life conditions are constraints on the selection, and other changes of the life conditions at other time points of the evolution could lead to an evolution that is completely different from the evolution of Earth. Let us denote this factor f_r and estimate that it may be 0.1.

If we introduce these three additional factors, for 1000 years, 10,000 years and 100,000 years of a technological civilization we derive that the number of other civilizations in the galaxy would be $0.33 \times 0.25 \times 0.1 = 0.008$ times what we previously found or respectively 0, 2 and 20 other advanced civilizations in the entire galaxy. Our galaxy has a diameter of 150,000 light years (Burnham et al., 2005). If we presume that there are two more technological civilizations in our galaxy and the distance is the reasonable one-third of the diameter, it is impossible to make any contact! The *Homo sapiens* could only make primitive stone tools 50,000 years ago. If we want physical contact, we should count on at least 500,000 years for a space ship to reach the other civilization, which makes the space travel impossible or meaningless. Maybe this revised interpretation of Drake's equation is a more reasonable guess? At least, it can explain why we have not yet received radio signals from other planets with civilization.

As the number of galaxies is enormous—probably in the order of 10^{11}, there is of course a very high probability that there is a couple of planets with advanced civilizations in each galaxy or let us say totally 2×10^{11} planets with technologically high civilization—but the distance is unfortunately so enormous that it will not be possible to make any kind of communication with planets in other galaxies. If there is a galaxy with a planet with advanced life as close as 1 million light years, it will take 2 million years to start a dialogue by radio waves—and by then our civilization has, with high probability, vanished. The enormous size of the universe is inconceivable.

Lately, it has been discussed that some of the best places to look for alternative life forms are ecologically isolated niches such as volcanic vents and deep caves. If it is assumed that life can readily emerge under the right environmental conditions, it is possible that life arose on Earth more than once (Davies, 2007). If we could find a second genesis by searching, for instance, for exotic microbes with a different or slightly different biochemistry, it could be considered a strong support for the hypothesis that life inevitably will emerge under the right environmental conditions, provided that the conditions prevail for sufficient long time.

Davies (2007) mentions four possible radical differences for a possible different biochemistry:

1. The amino acids are left-handed and the DNA is right-handed double helix for the life on Earth. But if life would start again from scratch, the amino acids could be right-handed and the DNA left-handed.
2. All organisms on Earth use the same 20 amino acids to construct the proteins. Alien microbes could use other amino acids.
3. Arsenic could replace phosphorus in the phosphorus-containing biochemical compounds of the known life forms. Arsenic is poisonous because it mimics phosphorus so well. Similarly, phosphorus would be poisonous to an arsenic-based organism.
4. Silicon could replace carbon because it has, as carbon, a valence number of four. Silicon can be arranged in rings and long chains as it is known for carbon. Silicon-based life would be the most radically different aliens and we would most probably not find them on Earth.

It has not yet been possible to find alien life forms on Earth, but would it be possible in our solar system? Could—it maybe another possibility for alien life forms—liquid ethane and methane on Saturn's largest moon, Titan, replace the role water is playing for the life forms on Earth?

6

From inorganic to poly-organic compounds

6.1 FORMATION OF SMALL ORGANIC MOLECULES

There is a rich literature about the evolution from small organic molecules, via simple cells and more and more complex cells, via poly-cellular organism, via fish, amphibian, reptiles and mammals to *Homo sapiens*. It is not the idea in the coming sections to repeat a description of all the evolutionary details and a description of the many fossils that are illustrating the evolution, but the idea is to indicate a few major characteristics by the most important steps of the evolution, and first of all to calculate the increase in the so-called β-value, the increase in the eco-exergy density and the increase in the eco-exergy flow rate, which are associated with the most important steps of the evolution.

This book is based on the ideas to give a thermodynamic description and explanation of the evolution, as also indicated by the title. It is therefore obvious that the focus will be on the above-mentioned three thermodynamic variables, although some of the most significant changes in the properties of the organisms over time will be mentioned to conceive why the increase in the selected thermodynamic variables have been major or minor. The scope is to attempt to present a thermodynamic interpretation of the evolution.

An interesting illustration of the creation of organization (dissipative structure) as a result of an energy flow through ecosystems concerns the possibilities to form organic matter from the inorganic components which were present in the primeval atmosphere. This is the first step in the long evolution. Since 1897 many simulation experiments have been performed to explain how the first organic matter was formed on Earth from inorganic matter. All of them point to the conclusion that energy interacts with a mixture of gases to form a large set of randomly synthesized organic compounds. Most interesting is perhaps the experiment performed by Stanley Miller and Harold Urey at the University of Chicago in 1953, because it showed that amino acids and fragments of DNA, mainly amino bases, can be formed by sparking a mixture of CH_4, H_2O, NH_3 and H_2—corresponding approximately to the composition of the primeval atmosphere (see Figure 6.1).

The outer planets are still made up mainly of hydrogen, helium, methane, ammonia and to a lesser extent of water. It is most likely that the same chemicals are abundant elsewhere in the solar system, and therefore probably also in its four inner planets. It is

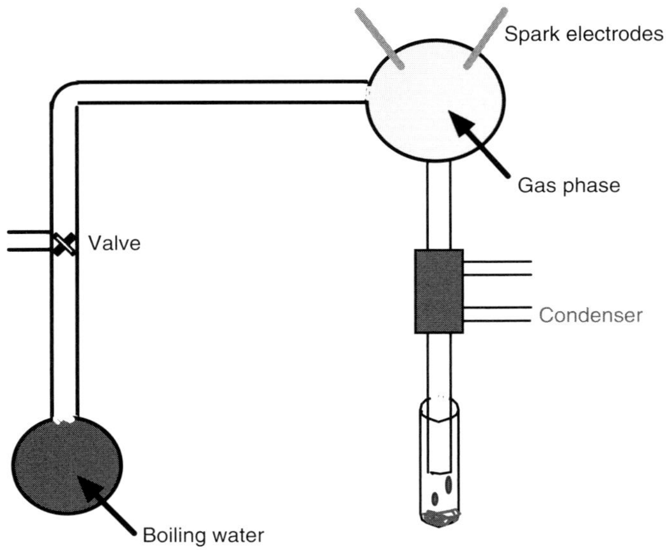

Figure 6.1 The apparatus used by Stanley Miller and Harold Urey to simulate reactions in the primeval atmosphere (reproduced from Jørgensen, 2002).

likely that the Earth's first atmosphere had great amounts of methane, ammonia and water and probably also a small amount of hydrogen. The chemical pathways from these simple molecules via, for instance, such compounds as hydrogen cyanide to amino acids and the amino bases are many and realistic. There is a rich chemical literature explaining the many possible pathways and how they can be combined to produce the entire spectrum of organic molecules that are the building blocks of the organisms.

It is remarkable that in a few days at the experiment performed by Stanley Miller and Harold Urey at the University of Chicago in 1953 about 20% of the gases were converted to amino acids and nucleotides. The same components are produced by the use of other energy sources and are detected in meteorites and in comets (Rasmussen, 2005). It is therefore, on the other side, not possible to decide whether the organic molecules were formed on the Earth or they were coming from the space. This is the so-called panspermia hypothesis, which addresses how biological material might have arrived on our planet and how life was originated in the first place. We will return to this question later in the book (Third Movement), when we have been able to obtain a thermodynamic overview of the entire evolution. No matter where it started, life had to arise from non-living matter.

The free energy content (kJ/mole) of the 20 amino acids that are used in the protein syntheses is clearly higher (about 1.5–2 times) than the free energy of other simple organic molecules that can be formed from a mixture of CH_4, H_2O, NH_3 and H_2, for instance ethane, ethylene, methylamine, but the solar radiation has anyhow sufficient energy to be able to produce the amino acids. The amino acids contain, in other words, more eco-exergy and they are formed although they require more free energy for the formation than several other possible simple organic molecules. These observations can be assumed a partial support for the tentative fourth law of thermodynamics (see Section 1.11).

Early, the primordial Earth was surely too hot for amino acids and nucleotide bases to survive. However, as soon as the Earth cooled enough for rocks to form and the atmosphere to be more protective, life arose. The oldest rocks solidified 4 billion years ago and the first life emerged about 3.8 billion years ago. It means that almost as soon as the temperature was sufficiently low, the amino acids and the nucleotides were formed—or coming from the space. It is noticeable that not only the 20 amino acids that are characteristic for the biochemical processes of organisms but also other amino acids are formed by an intense energy through-flow or are detected in meteorites. The 20 amino acids are however selected for the life processes, because they fit into the DNA and RNA coding (see also next chapter).

The polymerization of amino acids is not favoured in water. By heating a mixture of amino acids in the absence of water, it is however possible to aggregate amino acids into macromolecules which can even reach large dimensions (Barbieri, 2003), called proteinoids. They are not real proteins, because the amino acids are not arranged in linear chains, but form a variety of three-dimensional bonds. The proteinoids could have been the preliminary stage of proteins, as they are able to generate higher structures.

6.2 FORMATION OF POLYMER ORGANIC MOLECULES

The fuel of ecosystems is organic matter, detritus. It is therefore relevant to calculate the free energy of dead organic matter, consisting of poly-organic molecules. The chemical potential of dead organic matter, indexed $i = 1$, can be expressed from classical thermodynamics (e.g., Russel and Adebiyi, 1993) as:

$$\mu_1 = \mu_1^{\mathrm{o}} + RT \ln c_1/c_{1,\mathrm{o}} \tag{6.1}$$

where μ_1 is the chemical potential. The difference $\mu_1 - \mu_1^{\mathrm{o}}$ is known for detritus organic matter, which is a mixture of carbohydrates, fats and proteins. Approximately 18.7 kJ/g may be applied for the free energy content of average detritus. Obviously, the value is higher (22–24 kJ/g) for detritus originated from birds, as they in average contain more fat. Coal has a free energy content of about 30 kJ/g and mineral oil of 42 kJ/g. Both coal and mineral oil are a concentrated form of detritus from previous periods of the Earth. c_1 is the concentration of the detritus in the considered ecosystem and $c_{1,\mathrm{o}}$ is the concentration of detritus in the same ecosystem but at thermodynamic equilibrium.

Generally, $c_{1,\mathrm{o}}$ can be calculated from the definition of the probability, $P_{i,\mathrm{o}}$, of finding component i at thermodynamic equilibrium, which is

$$P_{i,\mathrm{o}} = c_{i,\mathrm{o}} \Big/ \sum_{i=0}^{n} c_{i,\mathrm{o}} \tag{6.2}$$

If this probability can be determined, then in effect the ratio of $c_{i,\text{eq}}$ to the total concentration is also determined. As the inorganic component, c_0, is very dominant at thermodynamic equilibrium, Equation 6.2 can be approximated as

$$P_{i,\text{o}} = c_{i,\text{o}}/c_{0,\text{o}} \tag{6.3}$$

By a combination of Equations 6.1 and 6.2, we get

$$P_{1,\text{o}} = [c_1/c_{0,\text{o}}]\exp[-(\mu_1 - \mu_1^{\text{o}})/RT] \tag{6.4}$$

The equilibrium constant for the process describing the aerobic (presence of oxygen) decomposition of detritus at 300 K can be found based upon the above-mentioned values. We could presume a molecular weight of about 100,000 (more accurate 104,400) (Morowitz, 1968) for the typical poly-organic molecules that make up dead organic matter, and with a typical composition of 3500 carbon, 6000 hydrogen, 3000 oxygen and 600 nitrogen:

$$C_{3500}H_{6000}O_{3000}N_{600} + 4250\ O_2 \rightarrow 3500\ CO_2 + 2700\ H_2O + 600\ NO_3^- + 600H^+ \tag{6.5}$$

$$K = [CO_2]^{3500}[NO_3^-]^{600}[H^+]^{600}/[C_{3500}H_{6000}O_{3000}N_{600}][O_2]^{4250}$$

since water is omitted from the expression of K. We have that

$$-\Delta G = RT\ln K$$

$$-\Delta G = -18.7\,\text{kJ/g}\ 104{,}400\,\text{g/mole} = 1952\,\text{MJ/mole} = 8.2\,\text{J/mole}\ 300\ \ln \text{K},$$

which implies that $\ln K = 793{,}496$ or K is about 10^{344998}.

The equilibrium constant is, in other words, enormous. The spontaneous formation of detritus in the form of a compound with the molecular weight of about 100,000 has therefore a very low probability.

Figure 6.2 illustrates the free energy of various ecologically important nitrogen compounds. By a steady input of energy, it is possible to convert nitrate or ammonium stepwise to proteins—to go up the free-energy staircase in Figure 6.2 step by step. Proteins with the highest free energy are in the present ecosystems, the result of the photosynthesis or of the biochemical synthesis in heterotroph organisms.

Detritus has an eco-exergy density corresponding to 18.7 kJ/g. The distance to thermodynamic equilibrium is therefore 18.7 kJ/g, which is a high value as also indicated by the high K-value. While the eco-exergy flow density is higher on the Earth than on the Sun, the eco-exergy density on the Sun is of course higher (see Section 5.1) as the Sun contains a total eco-exergy amount corresponding to the nuclear processes for the entire life-time of the Sun. Even if we consider detritus with a low molecular weight

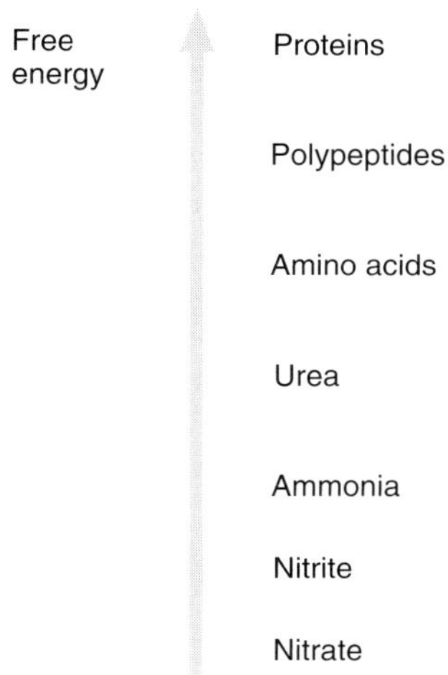

Figure 6.2 The level of free energy for some ecologically important nitrogen compounds.

corresponding to detritus partially decomposed, the *K*-value is still very high. If we presume a 100 times smaller molecular weight the exponent is 100 times smaller or about 3500—still a very high *K*-value. It is therefore understandable that detritus is decomposed spontaneously and thereby can yield energy to the heterotroph organisms. The opposite process corresponds to what may be the result of the photosynthesis, the conversion of solar radiation (energy) into chemical energy.

The formation of living cells in one step from non-living chemicals is impossible, and even the formation of the poly-organic molecules that are characteristic for life is impossible in one step according to the above shown calculations. The formation of poly-organic molecules must therefore have taken place stepwise, either by chemical non-biological processes or as a result of living organisms. The various components formed by an energy input to the primeval atmosphere can readily be adsorbed on clay particles or other fine particles. Amino acids have lower eco-exergy than detritus, for instance glycine 5 kJ/g, lysine 10.5 kJ/g and tryptophan 11.9 kJ/g. It means that the formation of the poly-molecules in living matter (detritus when it is non-living) requires an inflow of free energy. Further inflow of energy implies that the amino acids react to form polypeptides, first with a few amino acids, later with many more amino acids. Autocatalysis may also have played a role, when we are considering the formation of higher molecular proteins from polypeptides. It is not possible to determine whether the formation of poly-molecules took place before the first primitive life forms emerged or they were a consequent of the first primitive life forms (see also next chapter).

It is, however, interesting that concentrated solutions of proteinoids after heating to between 120 and 200 °C followed by a very slow cooling can form spontaneously vesicles, which Fox and Dose (1972) called microspheres. They have a regular form and with a diameter of 1–2 μm. They are very stable and retain the weak catalytic activity of individual proteinoids. They are also able to absorb proteinoids from the surrounding solution, whereby they can grow and divide into two by fission or budding. They also present a rudimentary type of metabolism. Still, however, the evolutionary potential of the microspheres remains a mystery. They might have appeared on the primitive Earth, but it is not sure if they had a future.

Another interesting theory about the first poly-molecules is the so-called surface metabolism. It is based on a solid thermodynamic argument. The formation of peptide bond is not favoured in solution, but on a surface. This is true for many polymerization processes. A great number of enzymatic reactions require collisions of three molecules, an event which is highly unlikely in space but much more probable on a surface. Generally, it is a thermodynamic principle that spontaneous reactions are more likely to occur on surfaces than in space or solutions. It is therefore reasonable to conclude that the first metabolic structures started out as two-dimensional systems, for instance on the surface of clay crystals.

7

From poly-organic molecules to eukaryote cells

7.1 THE FIRST STEPS TOWARDS A BIOCHEMISTRY

It cannot be excluded that the first primitive cells were coming from the space, but whatever the first biochemistry that was developed on the Earth or in the space, we should anyhow try to explain it. Developments over the past decade have given new credibility to the idea that Earth's biosphere could have arisen from an extraterrestrial seed. Planetary scientists have learned that our solar system has previously included many worlds with liquid water, which we consider the essential ingredient for the carbon-based life as we know it. Life may have evolved on Europa (Sweinsdottir, 1997), Jupiter's fourth largest moon, or on Saturn's biggest moon, Titan, which is rich in organic components. Biologists have furthermore discovered microorganisms durable enough to survive at least a short journey inside a meteorite.

Life processes are based on proteins with a certain and specific amino acid sequence that, it is necessary to remember, was able to start the evolution. We need therefore the genes and the proteins simultaneously, because the right amino acid sequence cannot be transferred to the next generation without the genes and the genes would have no function without the proteins that are controlling the life processes. We cannot imitate today the first biochemical processes—not yet at least—but a possibility could be that RNAs have been formed that should not be too difficult as fragments of these molecules are formed from inorganic molecules by an input of energy. RNA has an ability to replicate and to adsorb specific amino acids to the four amino bases that are the reactive parts of RNA. Probably these first contacts between RNA and amino acids have taken place on clay particles or at the surface of scum. RNA should therefore be able to build various polypeptides and they would contain the 20 amino acids that we know make up the life-determining proteins. In this context, it is important that a successful amino acid sequence, that is polypeptides or proteins with an amino acid sequence, which could serve as enzymes for useful conversions of other organic molecules, would be remembered. An obvious question would be: why are proteins made from 20 amino acids species that serve universally for their synthesis? All the amino acids that are used for the protein syntheses are, in addition, the L-forms—the D-forms are never applied. It is not a question about the relative abundance. A likely answer to this riddle is that the amino acids were selected for protein syntheses due to their ability to interact with RNA.

The first cells were much simpler than present-day cells. They are often denoted "protocells". They were subject to mutations. These affected the efficiency of the RNA, the resilience of the coding system and the quality of the synthesized proteins with respect to their ability to favour protocell growth and proliferation. De Duve (2002) presumes that the first proteins in an organism that may not even have a cell membrane were only 20 amino acids long, which implies a biological eco-exergy of $RT \ln 1/20^{-20} = 8.34 \times 300 \times 20 \times 3 = 150{,}120$ J/mole (see Equation 1.27, Chapter 1) or as the molecular weight of 20 amino acids is about 3000 g the eco-exergy in kJ/g becomes 0.05 kJ/g. This gives a β-value; see Table 1.1, Chapter 1, on $(18.7 + 0.05)/18.7 = 1.0027$ or less than the β-value for virus. The exergy density can be calculated to be $18.7 \times 1.0027 = 18.75$ kJ/g.

The proteins with a chain of 20 amino acids have 20^{20} or about 10^{26} possible arrangements. De Duve calculated (2002) that this amount of arrangements with even 99.9% of the volume to spare would only require a volume of $20 \times 50 \times 0.1 = 100\,\text{km}^3$. The experiment to find the best amino acid sequence takes place on the molecular level—nature uses super-nanotechnology. De Duve presumes that out of the 10^{26} different arrangements only 1000 were favourable in the sense that they were able to utilize organic matter for the metabolic processes and to favour growth and proliferation.

The next step is of course that the proteins react with each other and form proteins with bigger molecular weight. Virus can code for 2000 amino acids and organism similar to virus may have been the next step. Virus has a β-value of 1.01. Shortly (10^8 years?) after the Earth has got a suitable temperature on the surface, probably about 4.0 billion years ago, the life forms were as described above: a combination of RNAs and proteins that could reproduce and metabolize. The exergy density is $18.7 \times 1.01 = 18.9$ kJ/g.

7.2 THE PROKARYOTE CELLS

The three important steps from the very primitive life towards the prokaryote cell are as follows:

1. DNA was formed from RNA. DNA covers the information storage and replication while RNA utilizes the information protein synthesis. This division made it possible to regroup the genes and the replication of the genes could be carried out in a synchronous fashion, coordinated with cell division.
2. A cell membrane was formed to protect the life processes, the RNAs, DNA and proteins (enzymes). Physically, biological membranes are not very different from soap bubbles. They are very thin, highly flexible, self-sealing films, which consists of mainly carbon and hydrogen. Their constituent molecules are phospholipids. Formation of the cell membrane was a major progress. It became possible to protect the biochemical components making up the life processes and to maintain more easily another composition inside the cell than outside, but still mediate exchanges with the outside.

3. Introduction of ATP to facilitate the exchange of energy between different parts of the cell and between different biochemical processes. ATP is relatively easy to produce from other organic molecules by input of energy. It can therefore not be excluded that ATP participated in the biochemical processes of the very first primitive life consisting of small proteins and RNAs as described above.

The development corresponding to these three points has probably taken in the order of 100–200 million years. The oldest fossils of cells are about 3.8 billion years old and were found on Greenland (Haugaard Nielsen, 1999). The minimal cell has (see Table 1.1 and the comments to the minimal cell, Chapter 1) a β-value of about 5.0. The exergy density is now as high as $18.7 \times 4.88 = 91$ kJ/g. The free energy flow density of synthesizing organic polymer in primitive cells (Geigy, 1990) is in the order of 0.02 J/s kg. The eco-exergy flow density becomes therefore as it includes the accumulation of information $5 \times 0.02 = 0.1$ J/s kg. It is of course higher than for the Earth in average, as the microorganisms represent an intensive energy flow.

The primeval world was oxygen-free and remained so until about 2.3 billion years ago. The level of atmospheric oxygen started rising at that time due to introduction of the photosynthesis and reached values compatible with aerobic life a few hundred million years later. The organism responsible for this significant change in the composition of the atmosphere was *cyanobacteria*. Oxygen is deadly toxic to anaerobic organisms, but new life forms—aerobic microorganisms were able to cope with the new challenge of oxygen.

The time from about 3.5 to 2.0 billion years ago gave rise to a wide spectrum of different prokaryote cells using different biochemical processes, particularly to oxidize the organic matter, which is the energy source of these cells. Furthermore, the biochemical processes included in the metabolic processes became more refined. Some scientists talk about bacteria as superstars of the living world, because of the high reproduction rate. A β-value of 8.5 (see Table 1.1 in Chapter 1) represents the prokaryote cell in the most developed stage, which was approximately reached at the latest 2 billion years ago. The corresponding exergy density is $18.7 \times 8.5 = 159$ kJ/g. The eco-exergy flow density is about 0.02×8.5 J/kg s $= 0.17$ J/s kg, presuming the same turnover rate as for the primitive cell. The turnover rate is the same as measured today for some bacteria that can be found close to the so-called vents or black smokers.

7.3 THE EUKARYOTE CELLS EMERGE

The eukaryote cells are much larger than the prokaryotes. They are organism of considerable complexity. They are composed of many different parts, endowed with distinct function. They have a highly organized pulp, denoted cytoplasm, and a central kernel, called nucleus. Mitochondria contain the respiratory enzymes of the cells. Whereas the cytoplasm is the site of the majority of biochemical processes, including the protein synthesis, the nucleus takes care of the genetic operations. Eukaryotes have a well-organized system for duplicating their DNA exactly into two copies during cell division. This process, denoted mitosis, is much more complex and precise than the simple splitting found in prokaryotes. Eukaryotes can perform sexual reproduction in which the DNA of two cells is shuffled and re-dealt into new combinations. It has increased enormously the number of combinations that can be tested per unit of time for better fitness. The almost infinite variety possible from sexual reproduction provides for change in a changing world. Because of the constraints from changes of the life conditions, a successful genotype that would pass unaltered into the next generation would not be able to continue its success. Relying solely on an asexual reproduction, the capacity of species to adapt genetically is very limited; once its phenotypic flexibility has been exceeded there is little prospect of accommodating further changes. Sexual reproduction may be risky; but it does at least produce new combinations of genes and a possibility of increasing the range of tolerance. The cost of sexual reproduction is high, and so is the risk of failure. Each individual has to meet a partner and two gametes, the sperm and the egg must fuse to form a single cell, the zygote. The zygote must have a viable combination of genes, always with the risk that the new genotype may be unable to support the development of new individual. To survive, each offspring needs to be well matched to the environment in which it will grow and reproduce. When a large number of offspring are produced all are slightly different from each other and there is therefore an increased chance that some will survive in the changing world. There is, however, wastage and there is a major cost to the parents. Some combinations, some new genotypes will inevitably be less fitted than their parents. Asexual reproduction avoids all these risks and costs and may be more beneficial in a not changing environment. So, the constraints due to the changing life conditions is actually the prerequisite for the evolution (compare also with Chapter 2).

Table 7.1 The most significant differences between prokaryotes and eukaryotes

	Prokaryotes	Eukaryotes
Size	1–10 μm	10–100 μm
Nucleus	None. The chromosomal region is called nucleolus	Nucleus separated from cytoplasm by nuclear envelope
Intracellular organization	Normally, no membrane-separated compartments and no supportive intracellular framework	Distinct compartments, e.g., nucleus, cytosol with cytoskeleton, mitochondria, endoplasmic reticulum, Golgi complex, lysosomes, plastids
Gene structure	No introns; some polycistronic genes	Introns and exons
Cell division	Simple	Mitosis or meiosis
Ribosome	Large 50S subunit and small 30S subunit	Large 60S subunit and small 40S subunit
Reproduction	Parasexual recombination	Sexual recombination
Organization	Mostly single-cellular	Mostly multicellular, with cell differentiation

Source: After Klipp et al. (2005, p. 21).

The gradual accumulation of changes resulting from mutations and sexual recombinations and a few other mechanisms over many generations produce new species. The cells are surrounded with a great diversity of outer coverings of enormous complexity and made of a wide variety of substances, mostly proteins, carbohydrates, polymers of combinations of both. Finally, it should be mentioned that eukaryotes typically is 10 times larger in diameter or 1000 times larger in volume compared with prokaryotes. Had the transformation from prokaryotes to eukaryotes not taken place, the living world of today would still have been only bacteria.

Any textbook of microbiology gives a detailed description of the relatively significant differences between the prokaryote and the eukaryote cells. The differences are summarized in Table 7.1 The development from prokaryotes to eukaryotes started probably about 2.3 billion years ago, at the time when the oxygen concentration of the atmosphere started to rise (de Duve, 2002). Fossils of eukaryotes have been dated to 2100 million years ago (Madsen, 2006). The mitochondrion, which is the chemical factory of the eukaryote cells, was probably formed as a result of penetration of one cell into another, starting a biochemical cooperation.

There are strong evidences that the ancestors of eukaryotes were aerobic, although eukaryotes still may be divided into anaerobic and aerobic organisms, which indicates that the first phase in the development may have taken place before oxygen in the atmosphere started to rise (see the discussion in de Duve, 2002).

Typical early eukaryote cells are yeast and diatoms that represent the most developed life forms about 1.8 to 1.5 billion years ago. The step forward in the evolution from prokaryote to eukaryote is enormous and most probably archaea could be considered the link between prokaryote and eukaryote with β–values of about 13–14. Eukaryotes have β–values of 18–20. It implies exergy densities of about $18.7 \times 19 = 355$ kJ/g in average. The eco-exergy flow density becomes about 0.38 J/s kg.

8

Polycellular organization and the Cambrian Explosion

8.1 DIFFERENTIATION

Many bacterial colonies are known. They may be composed of several different (but often within the same genus) species that support each other by their specialization. The colonies followed four rules (principles) which can be considered a prerequisite for the further development towards poly-cellular organisms (Skøt, 2005):

1. All participants in the colony must have some sort of benefit as a result of the colony.
2. Communication among the components is a prerequisite.
3. The energy efficiencies and/or mass efficiencies are improved by the participation in colonies.
4. The colony offers better defence for the organisms.

What took a considerable amount of time was the development of the association from a single initial cell, which implies that all parts have the same genome. This new creative pathway started about 1 billion years ago. It is characterized by organism with cellular differentiation, that is the specialization in different directions of cells derived from the same parental cell. The differentiation into distinct cell types was possible because the cells do not express the totality of their genes but practise a selection that varies according to the cell type. Otherwise all the cells of an organism would be identical. The cells contain genetic switches that switch on and off the expression of certain genes, which is an important step towards cell specialization. The differentiation yielded a new advantage: division of labour. Different functions, such as digestion, biochemical processes, external protection, sexual reproduction, were carried out by different cells in the association. The cells are furthermore arranged into tissues and organs according to predetermined characteristics of the species, a disposition arising in the course of the embryonic development.

Another factor of importance for the evolution of pluricellular organisms is that they reproduce sexually, although also unicellular organisms as mentioned above sometimes join to reproduce. It implies that the new organisms arise from the fusion between two cells and therefore will have a genome corresponding to half of the genes from each of the original two cells. The main consequence of this system is a genetic diversification

Table 8.1 Composition of the atmosphere at different times in planetary evolution (Williams and da Silvia, 2003)

Time (years bp)	Composition of the atmosphere
4.5 billion	Methane, carbon monoxide, ammonia, hydrogen cyanide, hydrogen sulphide, nitrogen, hydrogen chloride
3.5 billion	Oxidized material as sulphate, very little oxygen, hydrogen peroxide, nitrogen
2.5 billion	Further oxidized material, modest oxygen, hydrogen peroxide, nitrogen
0.8–1.0 billion	About 10% oxygen, nitrogen
0.6 billion	As today or almost as today

within the same species, allowing it to adjust more rapidly and readily to the environmental conditions, which have an enormous variability (see Chapter 2).

About 1 billion years ago an association of unicellular green algae gave rise to the first primitive seaweeds. It is presumed that the oxygen production increased gradually about 800 million years ago extending the oxygen-bearing zone deeper and deeper into the ocean (Cowen, 2005). The oxygen content about 750 million years ago was probably about 10% of the present level. The oxygen concentration is extremely important for an effective use of the energy that is a prerequisite for the development of more advanced organisms. It created better conditions for more complex animals and plants to evolve. The evolution of the atmospheric composition is shown in Table 8.1.

The shift in the composition of the atmosphere has inevitably changed the available free composition of the sea, because the oxidation state of several elements has changed due to the increased redox potential. Many metals have very insoluble sulphide which was present in the early atmosphere and sea. As for the atmosphere, the sea is almost at steady state far from thermodynamic equilibrium, because its composition reflects the presence of living organisms. The most important changes in the sea from the original conditions to aerobic conditions are changes in the solubility of a number of metal ions. The very toxic heavy metal ions, mercury, cadmium and lead ions have, for instance, all increased their solubility in the sea due to a shift from anaerobic to aerobic conditions.

Sponges represent the type of organisms that emerged about 750–800 million years ago. They have a β-value of 98 corresponding to an exergy density of $98 \times 18.7 = 1833\,\text{kJ/g}$.

8.2 THE CAMBRIAN EXPLOSION

Slightly later, probably 600 to 500 million years ago, we are witness to what is called the Cambrian Explosion (the Cambrian period started 543 million years ago), because a surprisingly high number of new pluricellular organisms emerged. Fossils from as much as 50 phyla have been identified. The following fossils are characteristics for the Cambrian period (for β-values, see Table 1.1):

Jellyfish, β-value 91, exergy density $91 \times 18.7 = 1702$ kJ/g
Flatworm, β-value 120, exergy density $120 \times 18.7 = 2244$ kJ/g
Leeches, β-value 133, exergy density $133 \times 18.7 = 2487$ kJ/g
Roundworm, β-value 133, exergy density $133 \times 18.7 = 2487$ kJ/g.

To mention the organisms for which we have a β-value in the table. But many more organisms emerged: a tiny animal, *aldanella*, which may be the first mollusc, various types of sponges, lamp shell, trilobites, lopopods,*opabinia* with five (weak) eyes and one central grasping organ on its front.

From the fossil records and the corresponding β-values, it is possible to estimate that an eco-exergy density of about 1833 kJ/g has been reached about 750 million years ago, and an eco-exergy density of 2487 kJ/g represents what was reached about 540–530 million years ago, at the initial stage of the Cambrian Explosion. Peters (1983) indicates that the nitrogen flux rate for worms and leeches to synthesize proteins are about 0.02×10^{-6} g N/s and the amount of nitrogen is about 0.4 g, which gives a turnover rate of 0.5×10^{-7} l/s. The flux rate corresponds to an average size of worms and leeches, as the rate is size-dependent (see Peters, 1983). The eco-exergy flow density becomes therefore $1833 \times 1000 \times 1000 \times 0.5 \times 10^{-7}$ J/kg s $= 92$ J/kg s about 750 million years ago and $2487 \times 1000 \times 1000 \times 0.5 \times 10^{-7}$ J/kg s $= 124$ J/kg s about 540–530 million years ago—an enormous increase since the emergence of the eukaryotes, a factor of roughly 320 times.

Fossils of trilobites have also been found from the Cambrian period. Trilobites are arthropods with thick-jointed armour covering them from head to tail. They have antennae and large eyes. They were mobile on the seafloor using long-jointed legs and they were similar in structure to crustaceans and crabs. They represent therefore a major biological progress. The Cambrian and the Ordovician sea was populated with numerous trilobites in accordance to the fossil records. Probably about 60,000 different species of trilobites emerged as a result of the evolution. The extinction of the trilobites happened together with many other species by a catastrophic event in the Perm period. In spite of the extinction, they populated the Earth in almost 300 million years—compare with the genus homo that emerged about 2 million years ago.

The Cambrian Explosion was probably caused by a good combination of five factors:

1. *Higher oxygen levels*, which made it possible to increase very effectively the biochemical energy utilization (see Table 1.2 in Section 1.11). Large bodies and skeletons were made possible by high oxygen concentrations and shells and thick tissues prevent the free diffusion of oxygen into the body.
2. *Warmer weather*. The green house gas methane was removed from the atmosphere probably about 2000 million years ago by the increasing oxygen concentration. Up to about 600 million years ago, the Earth was covered completely by ice—the so-called snowball or slushball Earth. Many deposits about 750 to 600 million years ago

contain glacial debris even on equator, which indicate that the Earth was completely covered by ice at least during these 150 million years. The glacial sediments include dropstones, which fell from floating icebergs into the seafloor sediment. A volcanic emission of green house gases, particularly carbon dioxide, explains the warmer climate about 600–500 million years ago. The carbon dioxide could, in the beginning of this period, not be dissolved in the sea water due to the ice cover. The green house gases accumulated therefore rapidly in the atmosphere and forced thereby the temperature to increase and the ice to melt.

3. *Sufficient time* to allow to "survival test" of the various new pluricellular life forms.
4. *Introduction of predation.* Stanley (1973) has suggested that predation can cause additional diversity in their prey due to the development of various defence mechanisms. He argued that if predators first appeared in the early Cambrian, they may have caused the increase in diversity at that time. Predators may also have encouraged the evolution of many different types of skeletonized animals. The response of the preys to the emergence of the predators might be larger size, hard covering, powerful toxins or changes in lifestyle behaviour. As the new predators in turn evolve more sophisticated ways of attacking the preys, the responses and counter-responses might well have added to a significant burst of evolutionary changes. This is completely consistent with the role of constraints (see Chapter 2).
5. *A biotic enhancement factor for increased silicate weathering* implying a much faster cycling of silicates. This was significant for the formation of the skeletons that supported and protected the new Cambrian life forms.

The many species in the Cambrian period made it possible to develop food webs and complex ecological networks, which implied that the ecosystems moved further away from thermodynamic equilibrium (see Section 1.10).

The Cambrian explosion illustrates clearly that we have two directions of evolution: a vertical evolution, where genetic changes lead to more and more complex organisms, and a horizontal evolution, where the genetic changes are minor and where the body plan basically is not modified; it is variation on the same theme. The horizontal evolution is responsible for the enormous biodiversity. It is caused by the enormous variations in time and space (see Section 2.7). Due to the enormous variability in life conditions, small modifications will be favourable but dependent on the time and space. A typical example is Darwin finches. Darwin observed that on each island of the Galapagos Islands, the finches had differently shaped and sized beaks, adapted to the slightly different conditions and the slightly different foods on the different islands. Innumerable such examples could be summoned.

The vertical, in contrast to the horizontal, evolution corresponds to creation of new themes. It involves an increase in complexity due to a more stringent inner constraint. It illustrates clearly Monod's combination of necessity and chance. By a chance a new favourable combination of properties emerges and as it is absolutely necessary to survive by fitting to the steadily changed life conditions, the new combination will survive only if the combination of properties is better fitted to the prevailing conditions than other possible "solutions". The question is whether the number of mutations that should drive the vertical evolution is sufficient? The frequency of one wrongly inserted base by a replication error is 1 in 1 billion, but the probability after 34 generations of finding a new and perhaps favourable mutation is 99.9% (de Duve, 2002). That means less than one day for bacteria and about one month for animal cells! These numbers repeat our previous argument that the evolution can be explained by natural processes.

The Cambrian Explosion is characterized by both horizontal and vertical evolution. Radically new “solutions” (species, families) emerged and at the same time many similar but still slightly different “solutions” were offered by the evolutionary process—the biodiversity increased thereby significantly.

At the end of the Cambrian period and at the beginning of the Ordovician period it is assumed that gastropods, sea squirts, bivalves, brachiopods (lamp-shells) and molluscs have emerged on the arena (Cowen, 2005). They have a β-value from about 190 to about 310. It means that the exergy density has reached the level of $310 \times 18.7 = 5797$ kJ/g approximately 480 million years ago—a significant increase, more than a doubling, during the Cambrian period, which reflects the significance of the Cambrian Explosion. Peters (1983) indicates that the nitrogen flux rate used for protein synthesis for an average size of sea squirt and bivalves is about 0.03×10^{-6} g N/s and the amount of nitrogen is about 1.0 g, which give a turnover rate of 3×10^{-8} l/s. The eco-exergy flow density becomes therefore approximately 174 J/kg s. The eco-exergy density has, from the eukaryote cell, increased by a factor of 16 and the eco-exergy flow density has increased by a factor of about 450.

9

From the Ordovician period to the Jurassic period

9.1 FISH, VASCULAR PLANTS INCLUDING FERNS AND INSECTS

Near the end of the Ordovician time about 450 million years ago, a mass extinction took place. It seems to be closely linked with major climatic changes and not with volcanic eruptions or with an extra-terrestrial impact. A first pulse of extinction happened at the beginning of a big ice age and the second pulse occurred as it ended. Mass extinctions remove many organisms from the ecosphere, a situation that provides a major opportunity for the surviving organisms to evolve to fill out the empty ecological niches. Mass extinction removes the incumbent effect, which is valid in ecosystems as well as in social systems. The incumbent effect makes it difficult for newcomers to compete as already touched on in Section 2.1. By the mass extinction, the surviving creatures get an opportunity for major evolutionary innovations.

The earliest fish with hard parts did not have a skeleton, but mineralized bony plates covering their body. They emerged in the Ordovician time and are named ostracoderm ("plated skin"). The next step was heterostracan fish from the Silurian and Devonian time. They had a more efficient (stream-lined) shape and therefore more powerful swimming and a better manoeuvrability. They had a strongly plated head shield. With the next step, named osteostracans, the paired fins evolved, which improved the swimming and the manoeuvrability, further. Stepwise evolved the gills and the jaws. Sharks and rays emerged about 400 million years ago and had cartilaginous skeletons rather than bones. Sharks have excellent vision and smell and an electrical sense, which equip them nicely for hunting. The bony fish evolved simultaneously very fast, resulting in a lightening of the bony skeleton, increasing sophistication and variation in shape and the paired fins. Even flying fishes had evolved in the Triassic time, about 250 million years ago (for the time division, see Table 9.1) In the Carboniferous time about 300 million years ago, the lungfishes emerged on the ecological arena.

The β-value for fish is 499 (see Table 1.1 in Chapter 1), corresponding to an eco-exergy density of 9331 kJ/g. Obviously, the first more primitive fish should most probably be given a lower β-value. It is hard to give any more exact value than an increase from about 330 (6171 kJ/g) for the Ordovician time, about 450 million years ago, to 400 (7480 kJ/g) for the Devonian time, about 400 million years ago, and the above-mentioned 499 from the Carboniferous time, about 350–300 million years ago.

Table 9.1 Time division

Time	Million years ago
Precambrian	4550–543
Archean Eon	3800–2500
Proterozoic Eon	2500–543
Paleozoic	543–250
Cambrian	543–490
Ordovician	490–443
Silurian	443–417
Devonian	417–354
Carboniferous	354–290
Permian	290–250
Mesozoic	250–65
Triassic	250–206
Jurassic	206–144
Cretaceous	144–65
Cenozoic	65–0
Paleocene	65–55
Eocene	55–34
Oligocene	34–24
Miocene	24–5
Pliocene	5–1.8
Pleistocene	1.8–0.01
Holocene	0.01–0

With an approximate turnover rate of nitrogen on 0.4×10^{-7} l/s for a fish based on a size of 25–50 g, the eco-exergy flow density became about 247 J/kg s 450 million years ago, about 299 J/kg s 400 million years ago and about 373 J/kg s 350 million years ago.

The plants invaded the land during the Devonian time. From the Devonian time we know also the first tree, *Archaopteris*. The evolution of the plants was, however, slow, probably because plants are more vulnerable to competition than animals. Only few fossils are known from the Devonian time, while late Carboniferous time has been intensely studied, which has given us a good picture of the flora and global paleoecology about 320 million years ago. Swamp forests in tropical lowlands have dominated the landscape. Fern trees formed a dense understory several metres high under the 30 m canopy forest. The vegetation and organic debris that were deposited in oxygen-poor water formed thick accumulations of peat, now compressed as a giant coal belt. The vascular plants had lignin, which made them invulnerable to potential herbivores. That may have limited the development of plant diversity (see the discussion in Chapter 4, due to predation). But eventually, of course, both invertebrates and vertebrates developed herbivory. Bacteria and fungi can break down the toxins in dead plants and possibly a symbiosis made it possible for animals to eat living plant materials for enzyme-assisted digestion.

Plants have another pattern for the development of β-values. About 350 million years ago, they achieved a β-value of only about 158.

The first insects are Devonian, while the explosive radiation of winged insects happened in the Carboniferous time. Gigantic dragonflies with a wing span of 60 cm—the largest flying insects ever—were flying in the swamp forest. They were predators on smaller arthropods. Numerous fossil arthropods are recorded from the Carboniferous and Permian time—from about 325 to 250 million years ago. In this period the β-value for insects may have developed from about 160 to about 320—a doubling.

9.2 THE FIRST TETRAPODS AND AMPHIBIANS

It may be energetically cheaper and easier to extract oxygen from air due to the very low solubility of oxygen in water, particularly in warm water, and due to the higher density of water. Oxygen diffuses 100,000 times faster in air than in water. Oxygen-poor waters have often, however, a rich food supply. Because of the advantages of air breathing, many early fish did that probably. Once air breathing has evolved it can dominate respiration.

The Osteolepiforms, Devonian fish, that had lungs and used air breathing took the pathways towards tetrapods. Osteolepiforms improved their locomotion in shallow waters and mud-banks by development of stronger fins. It was a beneficial development because of the richer food supply in shallow waters and mud-banks. The fins gradually became fewer and stronger until toed feet evolved, which allow them to exploit even the shoreline of the shallow waters. Of course, many other changes (see Cowen, 2005) took place slowly but certainly over a period of million of years.

The fossils from the first tetrapods are from late Devonian time—about 360 million years ago. These early tetrapods spent their adult life in water with only occasional journeys on land mainly for spawning and basking. They were the first animals to exploit the water's edges. Later tetrapods (the Carboniferous time) adapt to different habitats and some became dominantly terrestrial while some others became aquatic and some became genuinely amphibians. The early tetrapods have still living descendents, which are sharply divided into amphibians and amniotes (reptiles, birds and mammals). Living amphibians are all relatively small-bodied and soft-skinned, and in this respect they are quite unlike early tetrapods. They are the salamanders, the frogs and the toads. Similar types of amphibians did not evolve until late Permian and even early Triassic time. There is not much guidance to the origin of these amphibians during the Devonian and the Carboniferous time. Fossils have been found, however, of *Eryops*, heavily built animals, much like a crocodile in size and probably also in ecology. They spent most of their time in water, as they were clumsy on land.

Amphibians have solved most of their problems associated with exposure to air, although their reproductive system is linked to water. Almost all amphibians spawn in water, where they lay a great number of small eggs that hatch quickly to swimming larvae. The eggs do not need complex protection against drying, because if the environment dries, the larvae as well as the eggs are doomed. The selection has therefore acted towards the efficient choice of suitable sites for laying eggs, rather than devices to protect the eggs.

The β-value for characteristic amphibians is approximately 688 corresponding to 12,866 kJ/g. A characteristic turnover rate of nitrogen used for the protein synthesis for a frog is 4×10^{-8} l/s, which means that the eco-exergy flow density becomes 514 J/kg s. These values correspond probably to the achievements of the evolution about 270 million years ago.

9.3 THE EMERGENCE OF AMNIOTES

Fish and amphibians may migrate long distance for spawning in water. Reptiles and the early amniotes had a different system. Their juveniles hatch into air as competent terrestrial animals. The stages of the embryological development of reptiles are similar to those of amphibians, but the reptiles develop for a longer time inside the egg. It means that the eggs must be larger and contain more food. The embryo is nourished by a large yolk. The eggs of reptiles are covered by an outer shell made of calcareous material. The membrane and shell layers allow gas exchange with the air for the metabolism of the growing embryo, but they are also able to resist water loss. The shell gives the eggs strength and reduces the temperature changes and the risk of desiccation.

The radiation of amniotes was encouraged by living opportunities away from the water bodies. It means that they were exposed to lower humidity, solar radiation, colder nights and greater temperature fluctuations. They needed therefore some degree of temperature control and thermoregulation. The life away from water bodies demanded adaptation for dealing with seasons and with temperature and rain changes and changes in food supply. The challenge of the early amniotes was to deal with a more changeable and less predictable environment.

The biochemical processes are run by enzymes, which are sensitive to temperature. Enzymes work best at an optimum temperature. Reptiles take on the temperature of the environment. Their body can function over a quite a range of internal temperatures, but there is also an optimum temperature for the body function. The reptiles try to control it at that level by behavioural thermoregulation. It takes more energy to stay warm but it is also possible at higher temperatures to increase the hunting and foraging efficiency. The specific surface increases with decreasing size. Small reptiles therefore bask in the Sun or exercise violently to increase their body temperature.

Amniotes became the dominant animals in all terrestrial environments during the Permian and Triassic time. Numerous fossils from this time give us a picture of a great variety of animals, representing early reptiles. Both herbivores and carnivores and even omnivores animals occupied the terrestrial ecosystems. The herbivores animals developed a method to break down cellulose. They chew the plant material well and have fermenting bacteria as symbionts to aid digestion.

Generally, the development was towards bigger and bigger animals. The energetic role of the size will be further discussed in the next chapter. Many reptiles in the Triassic time had a body length of 2–3 m or more. Some had fancy thermoregulations as, for instance, the pelycosaurs, *Dimetrodon*. The bones were covered with tissue to form a huge vertical sail. The sail allowed the animal to warm up quickly in the morning and to reach a high body temperature close to ours. It must have lived in an environment where such thermoregulations were both required and possible.

Thousands of specimens of therapsids have been found from late Permian rocks in South Africa. Their thermoregulation must have been more developed than simple behavioural regulation. We know that mammals evolved from late therapsids in the late Triassic time. It is an open question if the Permian therapsids already had mammalian style of thermoregulation, with automatic internal control and with a high metabolic rate.

A significant development took place from almost about 265 million years ago to about the start of the Jurassic time, a period of about 59 million years. Amniotes made many evolutionary progresses, associated with the respiration, metabolism and locomotion, in this period. In the middle of the Triassic time, about 228 million years ago, the evolution had probably still not reached the β-values (see Table 1.1) that are valid for reptiles (833) according to a comparison of the amniotes fossils with the reptiles of today. Let us presume a value of 750 for the time 228 million years ago. It means that the eco-exergy density is about 14,025 kJ/g. The nitrogen turnover rate is of course dependent on the size, but if we consider a lizard-like animal, the so-called *Hylomonos*, which we know was characteristic in a more developed form (the early forms could be found in the Carboniferous time) for the Triassic time, the turnover rate is in the order of 4×10^{-8} l/s, which means that the eco-exergy flow density becomes 561 J/kg s.

10

The evolution from dinosaurs to birds and mammals

10.1 REALM OF DINOSAURS

Dinosaurs dominated land communities during the Jurassic and the Cretaceous time. Many small mammals have emerged, but the dinosaurs were somehow competitively superior to the mammals of the time and confined them to small body size and ecological insignificance.

The early dinosaurs were bipedal carnivores and evolved from small bipedal archosaurs. They appeared in the late Triassic time—about the same time as the first mammals. The many variations of dinosaurs came later, but all four major dinosaur groups had evolved by the end of the Triassic time or by the beginning of the Jurassic time.

Birds are highly evolved dinosaurs, which mean that the clade Dinosauria includes birds as well as non-flying dinosaurs. Birds are covered in Section 10.2.

The largest dinosaurs were more than ten times the weight of elephants, the largest land animal alive today. The literature is rich in covering all the spectacular variations of dinosaurs, from stegosaurus and troodons to tyrannosaurus. We shall not here attempt to give an overview of the enormous variations and the wide spectrum of different forms of dinosaurs, but we should mention a few of the evolutionary progresses that have justified that dinosaurs carry more information than the amphibians.

Major finds of fossilized eggs and nests have shown that the nests were carefully constructed bowls of sand and mud lined with vegetation. The eggs were laid or arranged in a neat pattern. Many of the nests were clustered together at regular close intervals, suggesting that they were communal breeding grounds. It is also clear that the nests have sometimes been reused and remodelled in successive seasons. There are also good evidences for a long-term parental care by dinosaurs. So, in terms of their posture and behaviour dinosaurs had complex social structure and intelligence needed to run a complex society.

Were the dinosaurs homeotherms? Were they endotherms? Did they produce an optimum body temperature by a high metabolic rate? Or were they ectotherms, controlling their body temperature dominantly by clever use of the sunlight and by a high activity level? These questions cannot be answered with any certainty.

Metabolic rate among living vertebrates falls with increasing size in any group of similar animals. The dinosaurs—particularly, if they were cold-blooded—had probably a lower turnover rate than the mammals.

It cannot be excluded that the β-value for dinosaurs 150 million years ago was higher than 833, which is used for reptiles of today. Let us assume 900 and the eco-exergy density becomes 16,830 kJ/g, and a nitrogen turnover rate for animals of the same size as frogs is 5×10^{-8} 1/s, which means that the eco-exergy flow density becomes 841 J/kg s.

10.2 BIRDS

Primitive insects are known from the Devonian time but flying insects are not found before the Carboniferous time, as insects radiated in the Carboniferous forest canopy. It took more than 50 million years for insects to develop the ability to fly.

Several living forest-dwelling vertebrates have involved parachuting flight using skin flaps as flight surfaces, for instance the flying frog. Fossils of gliding vertebrates have been found in the late Permian time. Fossils of Coelurosauravus have been found in Germany, Britain and Madagascar. These areas were tropical shores at the time. Coelurosauravus is an ordinary small diapsid reptile in the structure of its skull and body. It has the size of a small squirrel. The trunk is, however, dominated by twenty or more long, curving, lightly constructed rod-shaped bones that extended outwards and sideways from the body. They supported a skin membrane that was close to an ideal airfoil in shape, 30 cm across. It was probably used for gliding.

Pterosaurs are the best-known flying reptiles. They evolved in the Triassic time, probably from gliding reptiles. Pterosaurs have lightly built skeletons with air spaces in the bones. The fourth finger supported the wing membrane and was about 3 m long in the largest pterosaurs. They were the largest flying creatures ever to evolve and flourished for more than 140 million years ago. For further details about these interesting reptiles, see Cowen (2005).

Living birds are warm-blooded with an efficient thermoregulation that maintains a high body temperature. Birds breathe more efficiently than mammals, pumping air through their lungs rather than in and out. They have better long-distance vision than any other animals. They build sophisticated nests and they are second only to humans in their ability to create art objects.

The earliest known birds is *Archaeopteryx* from Jurassic rocks in Germany. It was immediately recognized as a fossil of a bird, because opposite to the reptiles it has feathers on its wings and tail. But without the feathers it looks very much like a small dinosaur. The evolution of birds was very rapid. A number of fossils of birds more developed of course than the *Coelurosauravus* have been found from the Cretaceous time. Most of these birds were shoreline habitats. Good fossils of diving birds have also been found from this period. Further development of birds took place during the Cenozoic time, although at a more reduced rate of evolution.

Birds have a β-value of 980 equal to an eco-exergy density of 18,326 kJ/g. The nitrogen turnover rate is much faster than for reptiles due to the constant high temperature in birds (see Peters, 1983). The protein synthesizing turnover rate is in the order of

5×10^{-7} l/s for a 25–50 g bird—a size that is comparable with the size of a frog and a lizard (see Chapter 9). The turnover rate is about 10 times faster than for dinosaurs (see also the discussion in Chapter 5). It means that the eco-exergy flow density, let us assume around 100 million years ago, has increased to 9164 J/kg s. It is significant more than for the dinosaurs—a factor 13, provided that our assumption that the dinosaurs were not completely homeotherms is correct.

10.3 THE MAMMALS

The mammals had almost no significance in the Mesozoic time. The mammals were small and rare. They were and are in many ways very different from the dinosaurs and the living reptiles. Mammals suckle their young; they are homeothermic and endothermic and they have hair and no scales. Their teeth are not replaced continuously during life as for reptiles but are replaced (milk teeth) only once. The mammal brain is enlarged, providing them with improved sensitivity to hearing, smell and touch. Lately, it has been found that pregnancy and motherhood change the structure of the female mammal's brain, making mothers attentive to their young and better at caring for them. Recent experiments have shown that mother rats outperform virgins in navigating mazes and capturing prey. In addition to motivating females towards caring for their offspring, the hormone-induced brain changes may enhance a mother rat's foraging abilities, giving her pups a better chance of survival. Moreover, these cognitive benefits have been shown to be long lasting, persisting until the mother rats enter old age. Although studies of this phenomenon have so far focused on rodents, it is likely that other mammals also gain long-lasting mental benefits from motherhood. It is obvious that a strong selection of improved motherhood properties inevitably will take place. It is therefore understandable that the described hormone-induced brain changes have evolved, although they are biochemically complex. They may, however, have been a significant advantage in the competition between mammals and dinosaurs.

There are many more differences if we go into further details (see any textbook in zoology). These differences did of course not arise overnight; but as already mentioned in Section 9.3, the first mammals evolved about 200 million years ago, which means that the characteristic features and properties of living mammals, which we also find in fossils from the Eocene and Oligocene time about 30–40 million years ago, took about 160 million years to develop. Fossils demonstrate clearly the gradual evolution from dinosaurs to mammals. But considering the many beneficial properties of mammals, particularly their better senses and that they were homeothermic and endothermic, it is a puzzle why they were so insignificant until 65 million years ago, when all dinosaurs, plesiosaurians, mesosaurs and pterosaurs became extinct.

Many plankton species and tropical invertebrates became extinct 65 million years ago and also many land plants were severely affected. The extinctions were worldwide and all the continents were affected. Many organisms, however, survived: mammals, insects, birds and flowering plants. In the governing theory, the mass extinction is explained by an impact or either an asteroid or a comet. An impact structure 180 km in diameter and about 65 million years old has been identified close the Yucatan peninsula of Mexico. The theory is supported by high iridium content in rocks laid down precisely 65 million years ago. Asteroids have a relatively high iridium concentration. The iridium layer has been identified in more than 100 places. Almost simultaneously, a giant volcanic eruption took place, and it has probably contributed significantly to the mass extinction, particularly if we assume that major and long-term climatic changes were a direct consequence of the volcanic eruption. The impact produced enormous amounts of sulfate aerosols in the atmosphere that acted as nucleation for acid rain. Rain with pH about 0 was probably produced. The direct effect is sufficient to suffocate many air breathers and destroy plant foliage and dissolved the shells of many marine organisms. The North American continent was absolutely devastated and the global catastrophe among land plants and surface plankton affected drastically all normal food chains. Small animals, for instance lizards and the mammals, burrowed and hibernated. They could still find nuts, seeds, insects and larvae. The catastrophe was very severe but not too severe, because so many creatures survived. More than a decimation of the dinosaurs and a far lower reduction in the number of small animals is sufficient to shift the competition for the resources. It is relatively easy to show by an ecological model that the survival of an organism is very sensitive to the initial value (the concentrations after the catastrophe), which can explain the sudden shift from dinosaurs to mammals.

It is not completely clear whether these hypotheses are correct, and we will probably never be able to tell with complete certainty what the right explanation is. Nevertheless, the evolution from 65 million years ago to about 25–30 million years ago gave rise to a very impressive and varied set of organisms. The mammals and the birds conquered the Earth, so to say. A number of orders among the mammals evolved during this period and filled all the ecological niches that were empty after the mass extinction of, mainly, dinosaurs.

This evolution is also reflected in the thermodynamic calculations. Thirty five million years ago, the β-value has reached 2127, corresponding to an eco-exergy density of 39,775 kJ/g. The protein synthesizing turnover rate is approximately the same as for birds of the same size, namely 5×10^{-7} l/s. It is a rate that is 10 times faster than for reptiles (and dinosaurs?) and 12.5 times faster than for fish and amphibians. It means that the eco-exergy flow density has increased to 19,887 J/kg s.

Parallel to the evolution of mammals, particularly the flowering plants evolved. The angiosperms were in middle to a great expansion in the late Cretaceous time and the expansion continued into the Paleocene and Eocene time. The β-value reached 393 (see Table 1.1). From the late Cretaceous time to the Paleocene and Eocene time, the evergreen woodland changed in many parts of the world to largely deciduous forests. Deciduous trees survived much better the boundary events from the Cretaceous to the Paleocene time than the evergreens did.

10.4 THE ROLE OF THE SIZE

Many species of dinosaurs had an impressive size; but maybe the horse offers the best-known illustration of the role of size. The history of horses is perhaps also the best-known fossil vertebrate group. In the early Eocene (50–55 million years ago), the smallest species was the size of a cat, but other species weighted up to 35 kg. The Oligocene horses, about 30 million years ago, were bigger, probably weighting up to about 50 kg. In the middle Miocene, about 17–18 million years ago, grazing horses of the size up to 100 kg were normal. Numerous fossils have shown that the weight reached 200 kg 5 million years ago and about 500 kg 20,000 years ago. This is according to Cope's Law, which states that organisms have the tendency to develop towards a bigger size. This rule is of course not always valid, for instance on small island where dwarf sizes may be dominant. Many factors have influence on the evolution of the size, for instance the total availability of food and the density of food. Cope's Law seems to assume that the food is almost unlimited with a high density, which may be the case for grass, the preferred food by horses. The question remains: Why is it beneficial to increase the size, when the food is abundant and at high density?

Figure 10.1 shows a model in the form of a STELLA diagram that has been used to answer this question. The model equations are shown in Table 10.1.

The model has been used to calculate the efficiency for different maximum weights. The heat loss is (see Peter, 1983) proportional to the weight in the exponent 0.75. The growth rate follows also the surface, but the growth rate is proportional to the weight in the exponent 0.67 (see the equations in Table 10.1). The results are shown in Table 10.2 and the conclusion is clear: the greater the maximum weight the better the eco-exergy efficiency.

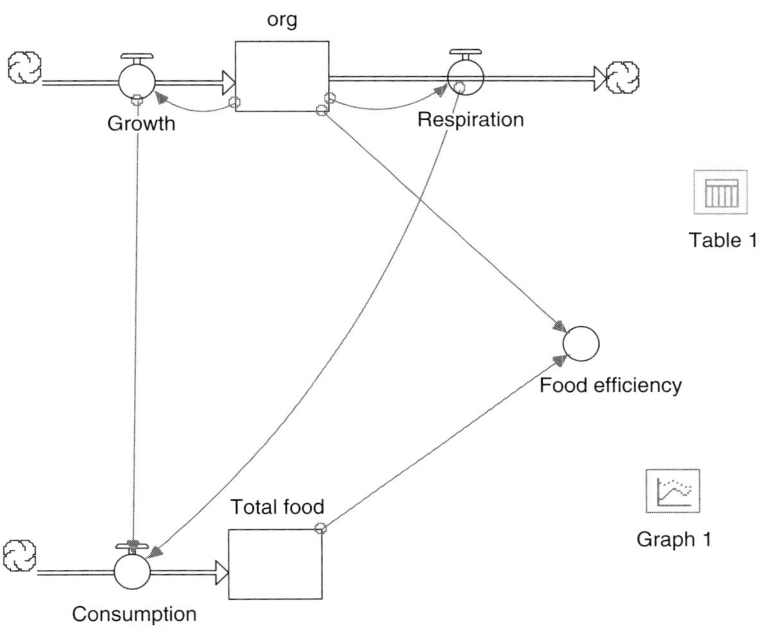

Figure 10.1 The growth and respiration is following allometric principles (see Peters, 1983). The growth equation is describing a logistic growth with a maximum weight. The food efficiency is found as a result of the entire life span, using the β-values for mammals and grass. The equations are shown in Table 10.1.

Table 10.1 Model equations (see the conceptual diagram, Figure 10.1)

d(org(t))/dt = (growth − respiration) (kg/24 h)
Growth = 3*org^(0.67)*(1 − org/upper maximum size) (kg/24 h)
Respiration = 0.5*org^(3/4) (kg/24 h)
d(total_food(t))/dt = (consumption) (kg/24 h)
Consumption = growth + respiration (kg/24 h)
food_eff % = 2127*100*org(t)/(200*total_food(t))
2127 is the β–value for mammals and 200 for grass

Table 10.2 Eco-exergy efficiency for the life span for different maximum sizes

Maximum size (kg)	Eco-exergy efficiency (%)	Upper maximum size parameter (kg)
35	1.41	45
50	1.55	65
100	1.84	132
200	2.20	268
500	2.75	690

The model results are not surprising, because a bigger size or weight means that the specific surface that determines the heat loss by respiration decreases. As the respiration loss is the direct loss of free energy, relatively more heat is lost when the body weight is smaller. Notice that the maximum size is smaller than the presumed upper maximum size that is a parameter to be used in the model equations (see also Table 10.2).

It should however be emphasized in this context that Cope's rule cannot be applied generally for all animals and plants. As discussed in Chapter 2, the selection pressure is a very complex concept and many other factors than just the efficiency of the food are determining the selection. There is a wide spectrum of constraints that are decisive for the evolution, including the presence of other species and the entire ecological network. Cope's rule and the model presented here should therefore be used carefully in other cases.

11

From primates to humans

11.1 FROM EARLY PRIMATES TO HOMINOIDS

The best-known early primates are plesiadapids from about 60 million years ago. They were ecologically like squirrels, had small brains and eyes. The teeth were adapted to cropping vegetation. They lived in North America and Europe. They looked and lived like rodents and may be considered a transition between squirrel and monkeys.

Various Eocene primates used most likely four-footed climbing and leaping from branch to branch. They used the full grasp of hands and feet for catching and holding small branches. The locomotion of lemurs, monkeys, gibbons, great apes and humans could have evolved from these general features of the early primates. Living anthropoids such are divided into three evolutionary groups: Old World monkeys, New World monkeys and hominoids (gibbons, apes and humans). Many fossils have shown the gradual evolution from the early primates to these three today living groups. Particularly, the emergence of hominoids has been studied intensely. More than 1000 fossils of hominoids from the early Miocene have been found. The best known is probably the *Proconsul* that weighted up to 35 kg. From a thermodynamic point of view, the evolution of the brain size is important, as the activity of the brain is significantly higher than the rest of the body—in the order of a factor 10. A gradual increase of the brain size is recorded from the early primates, to the *Proconsul*, to the gibbons, to the orangutans and to the chimpanzees.

If we compare with the fossils of early primates, it is reasonable to presume that a β-value of 2138 was reached about 30 million years ago. A β-value of 2145 corresponding to the anthropoid apes represents probably the evolution 10 million years ago if we use a single interpolation. The *Proconsul* (about 15–22 million years old) has probably a value between these two β-values. The corresponding eco-exergy densities are 39,981 kJ/g and 40,112 kJ/g and for *Proconsul*, 40,045 kJ/g. The eco-exergy flow rate for the early primates about 30 million years ago can be found from the turnover rate—5×10^{-7} l/s, the same as for other mammals. It yields an eco-exergy flow rate of 19,991 J/s kg for the early primates. For the hominoids about 10 million years ago, the turnover flow rate was higher due to the activities in general and due to the brain activity in particular. The activity has been measured in chimpanzee (Cowen, 2005) to be about 15 J/s kg. By multiplication with the β-value (see Table 1.1), we get an eco-exergy flow rate of 32,175 J/s kg. *Proconsul* had a brain volume of 170 cm^3 or about half of the chimpanzee's brain volume. An activity of 12.5 J/s kg can be estimated, corresponding to 26,083 J/s kg.

11.2 FROM ANTHROPOID APES TO HUMANS

The evolution from hominoids to humans has been intensely studied. Fossils for almost every step in this evolution have been found. Our lineage, the hominids, separated probably 6 million years ago (Lewin, 2005). *Orrorin* that dates 6 million years ago is considered the split between gorillas, chimpanzees and hominids. As many as six earlier species of homo ranging back to about 2 million years ago have become extinct (Cowan, 2005). Another six hominid species have been placed in the genus, *Austrolopithecus*, which goes back to about 4.4 million years ago and is accepted to contain the ancestors of homo.

The β-value has probably gradually increased from about 6 million years ago, where the value for anthropoid apes was valid, 2145, to about 120,000 years ago where the value for *Homo sapiens* can be used, 2173. There is almost 99% identity between the genes of the chimpanzee and *Homo sapiens*, but the main difference is between the regulatory genes (Jensen, 2004, 2005).

The brain size increased during the same period (Jensen, 2005; Lewin, 2005):

from 350 cm^3 for anthropoid apes, 6 million years ago,
to 385 cm^3 for *Australopithecus afarensis*, 4.4 million years ago,
to 450 cm^3 for *Australopithecus garhi*, 2.5 million years ago,
to 650 cm^3 for *Homo habilis*, about 2 million years ago,
to 900 cm^3 for *Homo ergaster* or *erectus*, about 1.4 million years ago,
to 1350 cm^3 for *Homo sapiens*, about 100,000 years ago.

Notice that the relative fastest growth of the brain size has been in the last 2 million years. From 6 million years ago to 2 million years ago, the bipedal walk was developed; but the brain did not grow much in size. A bigger brain made it possible to start to make tools and make a better hunting strategy that implied a more protein-rich food, which facilitated the growth of the brain.

The evolution has been much more complicated and for those interested in far more details, it is possible to refer to Lewin (2005). We can, however, use these six above-mentioned steps as milestones to obtain a thermodynamic indication of the evolution in the last 6 million years. We don't include the Neanderthals as milestone, because they lived at the same time as *Homo sapiens*. They had a more robust body and also a slightly bigger brain volume, namely 1450–1500 cm^3 (Jensen, 2005). The brain volume relative to the total weight was almost the same for *Homo sapiens* and Neanderthals. There are various hypotheses on why *Homo sapiens* was able to outcompete the Neanderthals, which are however not relevant for our calculations. The calculations are summarized in Table 11.1. The increase in the β-value has been assumed proportional to the brain size in these calculations. As the β-value is only about 1.5% higher for *Homo sapiens* than for anthropoid apes, the uncertainty introduced by this assumption is only minor.

In these calculations, it is furthermore assumed that the turnover rate increases from 15 J/s kg for anthropoid apes (see Section 11.1) to about 20 J/s kg for *Homo sapiens*, proportional to the brain size, as the increase in activity is due to the brain activity.

Table 11.1 The evolution in the last 6 million years, indicated by thermodynamic variables

Time (million years ago)	Organism	β-value	Eco-exergy density (kJ/g)	Eco-exergy flow rate (J/kg s)
6	Anthropoid apes	2145	40,112	32,175
4.4	*Australopithecus africanus*	2146	40,130	32,566
2.5	*Australopithecus garhi*	2148	40,168	33,294
2	*Homo habilis*	2153	40,261	35,525
1.4	*Homo erectus*	2160	40,392	38,340
0.1	*Homo sapiens*	2173	40,635	43,460

Climatic changes were, as mentioned above, the push that started the evolution from anthropoid apes to the genus homo via Australopithecus, but climatic changes have also made the evolution more difficult, for instance the colder and more arid climate about 60,000 years ago. The genetic variations of the present human population outside Africa indicate that there the human population passed a bottleneck about 60,000–70,000 years ago. Only about 2000–6000 individuals of *Homo sapiens* survived outside Africa (Jakobsen, 2005). The genetic variations in the human population today are far lower than the genetic variations in chimpanzees or gorillas. It is supported by Sykes (2001), who is able to relate the European population to seven daughters of Eve, as he formulates the reveals of our genetic ancestry.

Third Movement

A Holistic, Thermodynamic Interpretation of the Evolution

12

The three growth forms and the evolution

12.1 INTRODUCTION

The three growth forms can also be applied to interpret the evolution. All the three growth forms imply an increase of the eco-exergy as already presented in the First Movement. The Second Movement has been focusing on the increase of information and in this movement we shall summarize the results from the Second Movement and, in addition, we will present a wider thermodynamic interpretation of the evolution with particular emphasis on the evolution of networks. The evolution in the thermodynamic interpretation may be considered a steady increase of eco-exergy by growth of the biomass, the ecological network and the information.

The next chapter of this movement will look into the increase of biodiversity which is the prerequisite for a development/evolution of ecological networks, because the more components that we have available to build an ecological network, the easier it is to build the ecological network. The concept of ascendency is developed to account for the organization of network. We could therefore examine how the ascendency has developed during the evolution and compare this development with the development of information that was revealed in the Second Movement. It has, however, previously been found by the development of structurally dynamic models that the contribution coming from the energy-based ascendency, which is an excellent measure of the organization of ecological network, is minor compared with the contributions coming from the eco-exergy of the components in the ecological network. It will therefore be examined in this movement, Chapter 14, whether the use of eco-exergy to assess the flows among the components will be able to yield a higher ascendency that would better be able to match the eco-exergy of the components in the network. The results will be used in Chapter 15 to assess the evolution of the ascendency calculated for the most typical and probable networks. Chapter 16 summarizes the results of the Third Movement. Finally, the Coda will attempt to make a holistic summary and a conclusion of all the presented thermodynamic interpretations of the evolution. The Coda will try to see the forest through the trees.

In this chapter, we will, however, discuss the evolution of the first growth form, growth of the biomass. Probably earliest in the Devonian time, it has been possible in terrestrial ecosystems to capture the about 70–75% of the solar radiation that is physically possible. The question is therefore, how has the development of biomass been previously? This is the topic of the next section, which is based on the results of examination of aquatic ecosystems (Jørgensen, 2007a). The final conclusions will, however, be presented in the Coda. The evolutionary results of this examination attempt to cover the growth of biomass, the first growth form.

12.2 THE THREE GROWTH FORMS, APPLIED ON AQUATIC ECOSYSTEMS

Figure 12.1 illustrates how the three growth forms interact according to Figures 1.4 and 1.5 in Section 1.11. Growth form I is dominant in the first phase of the development from an early stage ecosystem to a mature ecosystem. By increasing the biomass, the solar radiation captured increases up to about 75%, corresponding to what is physically possible. Growth forms II and III are dominant in the intermediate phase and when the ecosystem is in a mature stage. Thereby more exergy is stored without increasing the exergy needed for maintenance. The system becomes, in other words, more effective in the use of the solar radiation according to Prigogine's minimum entropy principle. The exergy stored is increased for all the three growth forms.

The strategy ecosystems are using, when they are moving away from the thermodynamic equilibrium, seems very rational. Ecosystems do not have a strategy, but the concerted result of their properties explains these rational reactions. First, the ecosystem increases the physical-biological structure that as a parabola antenna captures as much solar radiation exergy as possible, that is the above-mentioned 70–75% that is physically possible.

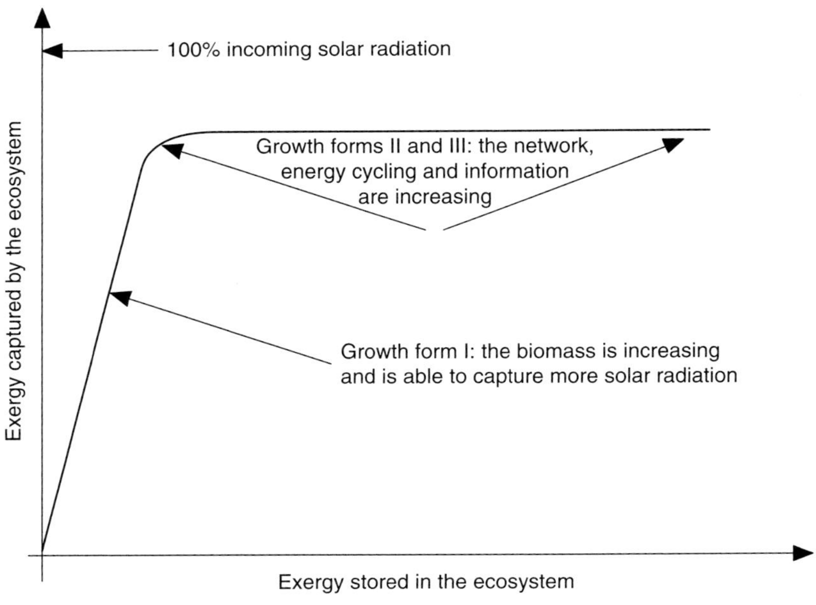

Figure 12.1 The development of an ecosystem is illustrated by plotting the exergy captured from the inflowing solar radiation towards the exergy stored in the ecosystem.

When the ecosystem is able to capture approximately as much exergy as it is physically possible, the ecosystem attempts to get as much out of the captured exergy as possible by increasing the efficiency. The network is increasing to recycle the energy more effectively.

The size of the organisms is also increasing to decrease the maintenance costs per unit of biomass, as the maintenance (respiration) is roughly proportional to the surface, that is to the length^2, and the biomass is of course proportional to the volume, that is to the length^3, which implies that the specific maintenance (respiration) cost becomes inversely proportional to the length. The regulation and the feedbacks of the organism become better and better and more and more sophisticated, as an ecosystem develops. The increased complexity of the organism and the higher content of information, which is characteristic of a mature ecosystem, will therefore imply that the efficiency of the energy use by the entire ecosystem increases. Notice that the growth of the network and the growth of the level of information do not require any resources, while the growth of the physical-biological structure requires the elements corresponding to the elementary composition. Growth forms II and III becomes in most cases dominant later as described above, although an overlap of the three growth forms takes place. The smaller the ecosystem is, which implies that it has a high relative openness, the faster will the growth forms II and III, particularly growth form III, contribute to the development of the ecosystem (Patricio et al., 2006). The recovery of small intertidal rocky communities has been examined in this chapter. It was found that biodiversity and specific eco-exergy (= eco-exergy/biomass) recover much faster than biomass and eco-exergy, that is that the growth forms II and III are dominant in the initial phase of recovery.

Aquatic ecosystems are expected to develop according to these considerations, that is follow the three growth forms and yield a Michaelis–Menten like graph when exergy destruction is plotted versus eco-exergy storage. Exergy destruction is in aquatic ecosystems expressed by the respiration, so the question is: what kind of plots is produced when respiration is plotted versus eco-exergy storage for different aquatic ecosystems? If a plot similar to Figure 12.1 is produced, it may be considered a support for the description of the ecosystem development by the three growth forms.

Christensen and Pauly (1993) have compiled a number of ecosystem representations giving information about the steady state concentrations of the most important organisms and the inflows and outflows for all the state variables, which are used to describe the ecosystems by steady state models. These data can be applied to calculate the eco-exergy of the ecosystem as described by the model and the total respiration rate of the ecosystems. The eco-exergy is usually found as

$$\text{Eco-exergy} = \sum_{i=1}^{i=n} \beta_i C_i, \quad (\text{g detritus equivalent/m}^2) \tag{12.1}$$

where β_i is a weighting factor of the ith species accounting for the complexity of the various organisms or the information that the organisms have embodied in their genes.

The respiration is in most cases indicated directly in g/m^2 yr or in a unit that can easily be converted to g/m^2 yr. The specific eco-exergy defined as eco-exergy divided by the biomass is found for all the 26 ecosystems that have been selected for this investigation. It should be expected that the ecosystems with relatively high eco-exergy are well-developed (mature) ecosystems, which entails that the specific eco-exergy is high due to a relative high concentration of developed organisms (with relatively high β-values).

The results (Jørgensen, 2007a) are shown in Table 12.1 and in graphic form in Figure 12.2. Different ecosystems have, however, different maximum respirations that can be found in the literature (see, for instance, Odum, 1969b). We can divide the examined aquatic ecosystems into three classes according to respiration rate: coral reef that can reach a respiration of 30 kg detritus/m^2 yr, fertile estuaries and lagoons that represent a respiration up to about 12 kg detritus/m^2 yr and fertile ponds and lakes that would be able to reach a respiration of about 4–5 kg/m^2 yr. Figure 12.3 shows the same results as in Figure 12.2, but the three levels of respiration are shown as horizontal lines on the figure. The types of ecosystems are also indicated on the figure for all ecosystems with more than 100,000 g detritus equivalent/m^2. It is clear from this interpretation of the results in Figure 12.3 that aquatic ecosystems have an upper limit for the amount of eco-exergy needed for maintenance, as it is also the case for terrestrial ecosystems.

Table 12.1 Respiration rate in g detritus equivalent/m^2 yr and eco-exergy storage in g detritus equivalent/m^2 for 26 aquatic ecosystems

Ecosystem	Respiration rate	Eco-exergy storage	Specific eco-exergy
Brunei, S. China Sea	1,346	7,378	47.7
Caribean coral reef	29,364	722,668	134
Coral reef, Philipines	24,290	313,251	108
Shelf ecosystem, Venezuela	4,149	17,805	69.0
Gulf of Mexico	1,615	11,147	41.1
Yucatan, Mexico	1,012	12,229	26.6
Etang de Thau, France	9,393	365,001	52.9
Maputo Bay, Mozambique	3,086	33,767	25.2
Campede Bank, Mexico	1,015	18,393	186
Gulf of Mexico	1,123	14,933	25.6
Lagoon, Vera Cruz, Mex.	1,262	5,049	14.9
Celestun Lagoon, Mexico	4,002	76,085	130
Tannahua, Mexico	532	4,717	151
Garonne River, France	2,447	8,551	150
Thames River, UK	1,458	27,025	352
Lake Chad	4,673	303,483	99.0
Lake Victoria	2,112	14,100	18.0
Lake Tanganyika	3,480	20,612	185
Lake George	3,216	18,331	183
Lake Turkana, Kenya	2,439	13,200	29.2
Lake Malawi/Nyasa	2,190	8,585	86.3
Lake Kinneret, Israel	1,930	7,900	34.1
Veli Lake, France	6,559	123,741	166
Laguna de Bay, Philippines	5,476	44,700	49.0
Pond in Malawi	2,848	30,450	175
French Frigate Shoals	1,200	23,391	52.9

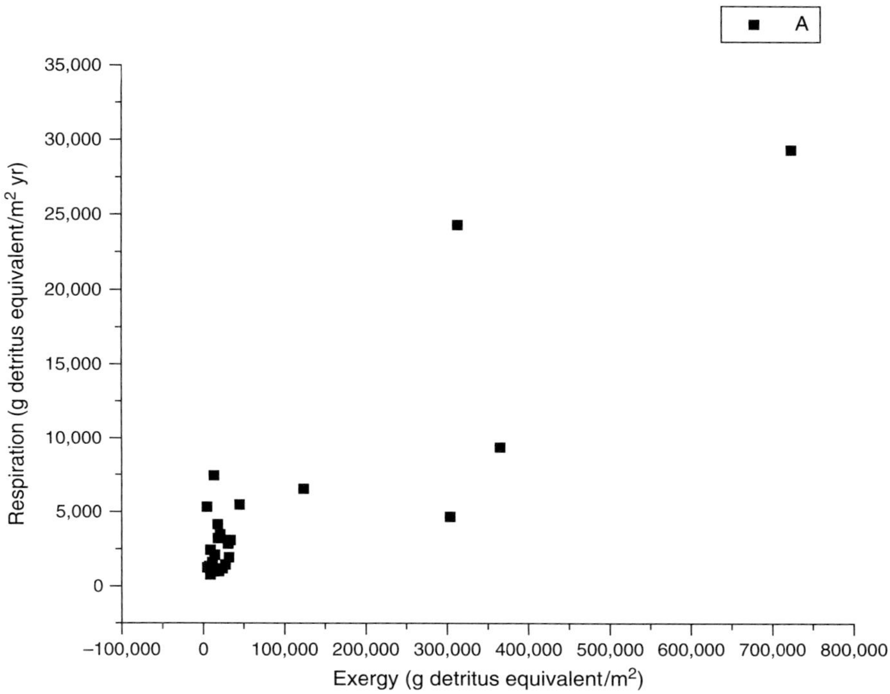

Figure 12.2 The eco-exergy storage expressed as g detritus equivalent/m^2 is plotted versus the eco-exergy used for maintenance (respiration) in g detritus equivalent/m^2 yr for 26 different ecosystems. By multiplication by 18.7, the results are obtained in kJ/m^2, as the average free energy (eco-exergy) of detritus is 18.7 kJ/g. The respiration is in most cases indicated directly in g/m^2 yr or in a unit that can easily be converted to g/m^2 yr.

The ecosystems that are represented in Figure 12.1 are presumed to develop towards an old forest system; but with possibilities to increase further the stored eco-exergy, that is move further away from the thermodynamic equilibrium by development of more complex ecological networks and a higher information content.

Arctic ecosystems, tundra ecosystems and mountain ecosystems above the timberline will, for instance, not be able to capture as much sunlight as a tropic rainforest or a well-developed temperate forest.

On the basis of the obtained results, it is assumed that ecosystems use the three growth forms to move away from the thermodynamic equilibrium. Increased biomass means that the ecosystem is able to capture more solar radiation. Increased biomass also means that more energy is needed for the maintenance. If we don't consider any other change than the biomass and that the size of the organisms are unchanged, twice as much biomass requires of course twice as much energy for its maintenance. Different climatic conditions determine which type of ecosystem that the climate on a long-term basis can sustain and different types of ecosystems will have a characteristic maximum biomass

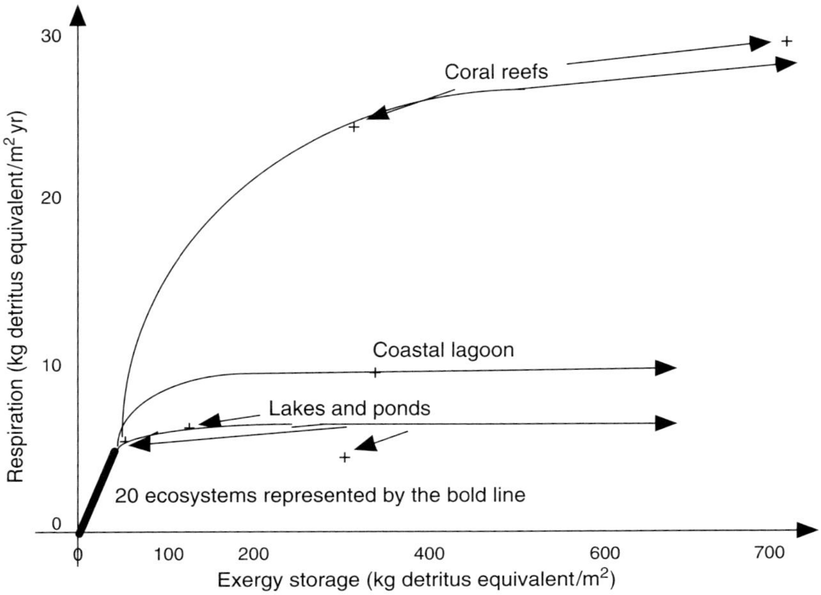

Figure 12.3 The figure shows the same results as in Figure 12.2, but three levels of respiration, namely for (1) coral reef, (2) fertile lagoons and estuaries and (3) fertile lakes and ponds, are shown as almost horizontal lines on the figure. The types of ecosystems are also indicated on the figure for all ecosystems with more than 44 kg detritus equivalent/m^2.

that they will be able to obtain (Odum, 1969a,b). Different types of ecosystems will therefore have a characteristic maximum maintenance energy demand. When an ecosystem has reached this characteristic level of maintenance free energy or eco-exergy and at the same time a characteristic level of solar energy that the ecosystem captures according to the biomass (mainly the vegetation), the ecosystem can still develop—move away from the thermodynamic equilibrium or increase the storage of eco-exergy—by the two other growth forms: develop the ecological network further and increase the information embodied in the system. The plots in Figures 12.1–12.3 are all supporting this interpretation of the three growth forms. All ecosystems seem to increase the biomass up to the feasible level, which is different for different types of ecosystems. There is no reason to believe that it has been different previously in the evolution. What has been different is of course to what extent it has been feasible for the pre-Cambrian ecosystems to utilize the first growth form—growth of biomass/physical structure. As the ecosystems today have different abilities to utilize the growth of biomass and physical structure, it would be obvious that the differences have been more pronounced only if we include in this comparison the pre-Cambrian ecosystems.

The specific eco-exergy for the 5 ecosystems with more than 100,000 g detritus equivalent/m^2 is in average 112, while the average for the 21 ecosystems that have less than 100,000 g detritus equivalent/m^2 is 94. The difference is, however, not statistical significance (a *t*-test was applied). The different ecosystems give very different conditions for fish and crustaceans which are the organisms that are contributing most to a high specific eco-exergy in the examined ecosystems. The two rivers included in the examination have a relatively high specific eco-exergy in spite of a relatively low eco-exergy due to a high fish concentration. If we, on the other hand, take the same types of ecosystems—and it is most obvious to examine lakes, because we have 10 lakes and ponds among the 26 examined ecosystems—we get a more clear picture. The 3 lakes with a lower eco-exergy than 15,000 have only a specific eco-exergy of 18, 29 and 34 (average 27 ± 11), while the 4 lakes with a specific eco-exergy of 100 or more (average 177 ± 15) have an eco-exergy of 18,000 or more. The difference between these two sets of specific eco-exergy is significant according to a *t*-test. The relationship between specific eco-exergy and eco-exergy storage is, however, not giving a good linear fit, which probably is due to variations among both the ecosystems and the modelling approaches. Anyhow, these results point towards the use of information as a growth form, in addition to the first growth form.

In the Coda, we will return to the interaction of the three growth forms and the role of biomass growth in the evolution of information and networks.

13

The evolution of diversity

13.1 THE ROLE OF DIVERSITY

Figure 2.2 has shown the evolution of an index combining diversity, expressed by the number of families in marine ecosystems and the information factor, β. The figure indicates an exponential increase.

The growth of ecological networks, denoted growth form II, is of course dependent on the diversity. The more different species, the better is the possibility to build a larger, more effective and better organized ecological network. It is therefore important, before the evolution of ecological networks is scrutinized, to examine how the diversity has evolved.

The relationship between biodiversity, measured by the number of species, and stability was previously widely discussed (see, for instance, May [1973], who showed that there is not a simple relationship between biodiversity and stability of ecosystems).

Tilman and his coworkers (Tilman and Downing, 1994) have shown that temperate grassland plots with more species have a greater resistance or buffer capacity to the effect of drought. The resistance or buffer capacity could be expressed by the change in biomass between a drought year and a normal year. However, there is a limit—each additional plant contributed less (see Figure 13.1).

Previously, it has been shown that for models there is a strong correlation between the eco-exergy and the sum of many different buffer capacities. Many experiments (Tilman and Downing, 1994) have also shown that higher biodiversity increases the biomass and therefore the eco-exergy, because by a higher biodiversity more ecological niches will inevitably exist and be utilized. There seems, in other words, to exist relationships between biodiversity and eco-exergy, between eco-exergy and resistance or buffer capacity and between biodiversity and buffer capacity.

It is possible from fossil records to estimate the evolutionary increase of the biodiversity and the species richness or what we could call the horizontal evolution. The results of the estimations are presented in the following section.

13.2 THE EVOLUTIONARY DEVELOPMENT OF THE BIODIVERSITY

Biodiversity may cover several diversity concepts as, for instance, the number of species, the number of families, the genetic diversity, Shannon's index, which may be considered just a logarithmic expression for the species diversity, the landscape diversity expressed by the number of biotic and abiotic constituents in the landscape and the number of

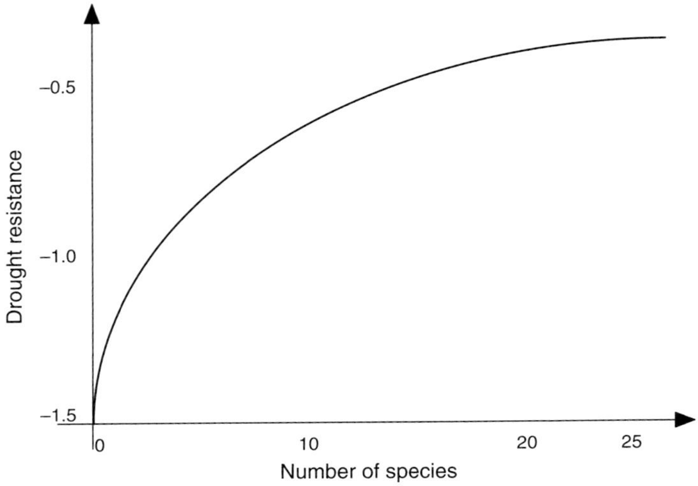

Figure 13.1 The results of the experiments in grassland by Tilman and Downing (1994). The higher the number of species, the higher the drought buffer capacity, although the gain per additional plant species is decreasing with the number of species.

different ecological niches. We will here use the number of species as the measure of the biodiversity, also denoted the species richness. A more revealing description of diversity can be undertaken by considering the distribution of individuals among species. It is possible to have a measure of diversity based on the combination of species richness and species relative abundance. It implies that the evenness in the form of an evenness index is considered. Biodiversity is a very inclusive concept, involving several facets and levels of organization (Dirzo and Mendoza, 2008). The development of new methods to analyse and characterize diversity is a very active field of research (for details, see Dirzo and Mendoza, 2008). Sometimes it is not possible to find the number of species, for instance, when the investigation is based on fossils. As we have seen, it is at least sometimes possible on the basis of fossils to determine with a reasonable accuracy the number of families, for instance the number of marine families.

Puzachenko (2006) has, based upon the data base by Benton (1993), proposed two statistical models for the increase of the number of families:

$$\log N = (-0.07861 + 0.031733 \log T)T \log T, \tag{13.1}$$

and

$$N = \exp[0.030053(1.03847T)]T \tag{13.2}$$

where N is the number of families and T is the time expressed by the unit 1 million years.

The two equations assume an exponential increase of the number of families and, as we assume, also of the number of species. Figures 13.2 and 13.3 show the number of marine families and insect families according to the fossil records (see, for instance, Dirzo and Mendoza, 2008).

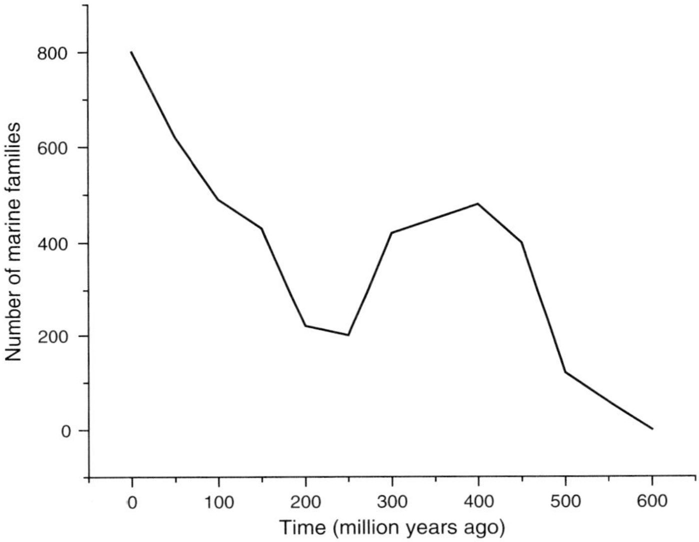

Figure 13.2 The number of marine families according to the fossil records.

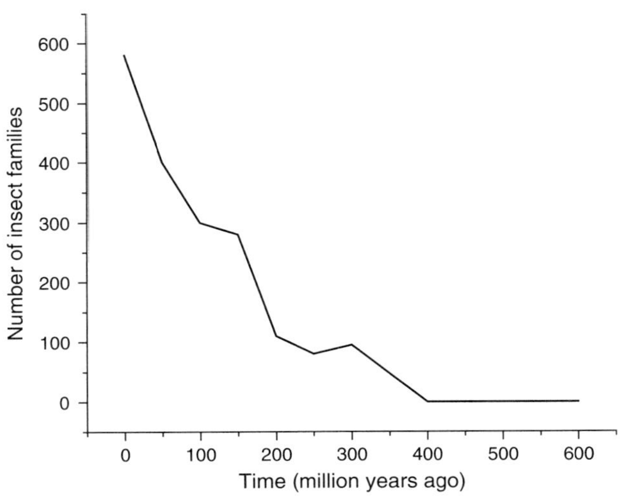

Figure 13.3 The number of insect families according to the fossil records.

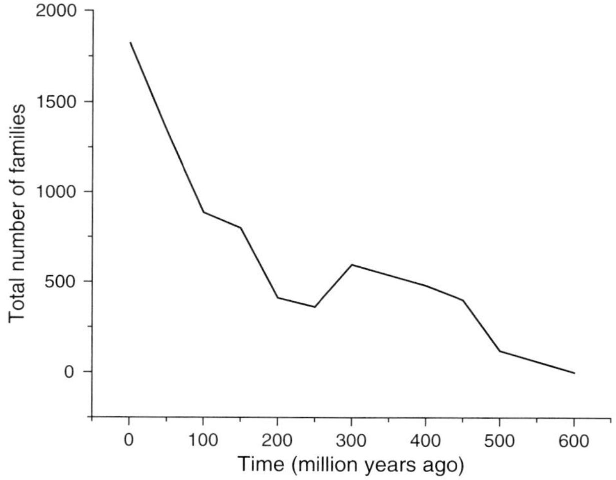

Figure 13.4 The number of families, totally, including marine families and insect families, according to the fossil records.

Figure 13.4 shows the total number of families including marine families and insect families. The figures show as the equations indicate an exponential increase of the species richness. The evolution has increased the species richness exponentially at least until *Homo sapiens* started to disturb nature about 5000–10,000 years ago. In the next two chapters, we will focus on the organization, size and efficiency of the ecological networks and ask the questions:

1. Has the increase in species richness been utilized by nature to increase also the organization, size and efficiency of the ecological networks?
2. Has the ascendency as a measure of the organization of networks increased parallel to the increase of the species richness?
3. Has the evolution of ecological networks implied a higher eco-exergy of the ecosystems?

Notice that Figures 12.2–12.4 show a decrease at about 250 and 200 million years ago, when two major climatic catastrophes occurred. The plots start all at $T = 0.1$, that is 10,000 years ago. It implies that the decline in species richness due to the development of agriculture and the decline in species richness that took place during the last 200–300 years due to the industrial development are not shown on the figures. When agriculture was developed about 8000–10,000 years ago, only a limited number of plant and animal species were utilized, of course. The development in the last 200–300 years has reduced the area of natural ecosystems and therefore the species richness.

14

Eco-exergy and ascendency

14.1 INTRODUCTION

It has previously been shown that the biomass or the energy content of the components in a network increases with the number of linkages between the components, because of an increased efficiency in the use of the available work energy (see Jørgensen et al., 2000). As eco-exergy is *RTBK*, where *R* is the gas constant, *T* the absolute temperature, *B* the biomass and *K* the Kullback's measure of information (see, for instance, Jørgensen et al., 2000, 2004, 2005; Jørgensen, 2002), eco-exergy will also increase with the number of linkages. The power, the eco-exergy storage and the ascendency follow the same trends, in accordance with Jørgensen (2002; see also Section 1.11; Fath et al., 2004; Ulanowicz et al., 2006). If the loss of energy by respiration decreases by increased size of the organism in the network (see Peters, 1983) or by increased information enables a better regulation, the biomass, energy, eco-exergy, power and ascendency all will increase, too, as mentioned in Section 1.11.

The calculations of the ascendency and power are generally based on energy, but it would be interesting to see what results we would get by using eco-exergy directly. Eco-exergy storage has been used as a goal function in structurally dynamic models (see Sections 1.11 and 3.1). As the network, however, also may change by significant changes of the forcing functions, it would in such cases be necessary to include ascendency or power in the goal function to be able to capture adaptations of the network to new prevailing conditions. The contribution from ascendency has, however, been relatively small compared with eco-exergy because the ascendency calculations were based on energy. It would therefore be interesting to see how much changes in the network would contribute to changes in the eco-exergy-based ascendency. In Chapter 15 the evolution of the network will be illustrated by the use of ascendency as the natural measure of the increasing organization of the networks during the evolution. A comparison of the evolution of the information (see Second Movement) and the evolution of the network would be interesting to carry out, but it would require that the same units and the methods of calculations are applied. A comparison by using ascendency calculated based on energy flows would inevitably show that the ascendency is almost negligible relative to the information based on eco-exergy. In addition, the flow of information is also important in ecological networks and it would therefore be natural to use eco-exergy when the evolution of networks and the evolution of information are compared or even added to get a total holistic image of the evolution.

Network, power and ascendency calculations based on eco-exergy have therefore been carried out and compared with the previous energy-based calculations.

14.2 ASCENDENCY AND CONNECTIONS

Ascendency represents the information stored in the network and the definition is shown below.

In more quantitative terms, the formula for ascendency is

$$A = \sum_{i,j} T_{ij} \log\left(\frac{T_{ij}T_{..}}{T_{i.}T_{.j}}\right).$$

where T_{ij} is the flow from compartment i to compartment j. T_i is compartment i and T_j is compartment j. T is the sum of all flows.

Figure 14.1 is used to illustrate the calculations of ascendency. The example is taken from Jørgensen et al. (2007). The calculations are presented in detail below the figure.

In Figure 14.1 is depicted the energy exchanges (kcal $\mathrm{m}^{-2}\,\mathrm{yr}^{-1}$) among the five major compartments of the Cone Spring ecosystem. The total system throughflow (TST) of Cone Spring is simply the sum of all the arrows appearing in the diagram. Systematically, this is calculated as follows:

$$\begin{aligned}\mathrm{TST} &= T_{01} + T_{02} + T_{12} + T_{16} + T_{17} + T_{23} + T_{24} + T_{26} + T_{27} + T_{32} + T_{34} + T_{36} \\ &\quad + T_{37} + T_{42} + T_{45} + T_{47} + T_{52} + T_{57} \\ &= 11,184 + 635 + 8881 + 300 + 2003 + 5205 + 2309 + 860 + 3109 + 1600 \\ &\quad + 75 + 255 + 3275 + 200 + 370 + 1814 + 167 + 203 \\ &= 42,445\ \mathrm{kcal\,m^{-2}\,yr^{-1}},\end{aligned}$$

where the subscript 0 represents the external environment as a source, 6 denotes the external environment as a receiver of useful exports, and 7 signifies the external environment as a sink for dissipation.

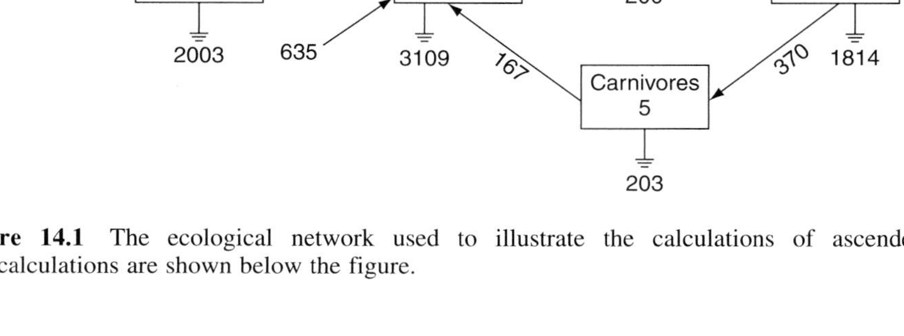

Figure 14.1 The ecological network used to illustrate the calculations of ascendency. The calculations are shown below the figure.

The role of connections is illustrated in Figure 14.2, presuming steady state (input = output) and first-order donor determined flows, which is often used in modelling of ecological networks. The examples are taken from Jørgensen et al. (2007).

Figure 14.2a shows the through-flow and exergy storage (based on a retention time of 5 time units) in the two components with no coupling, that is no network connections. Making a simple connection between the two links them physically, and while it changes their individualistic behaviour, it also alters the overall system performance. In this case, the through-flow and exergy storage increase because the part of the flow that previously

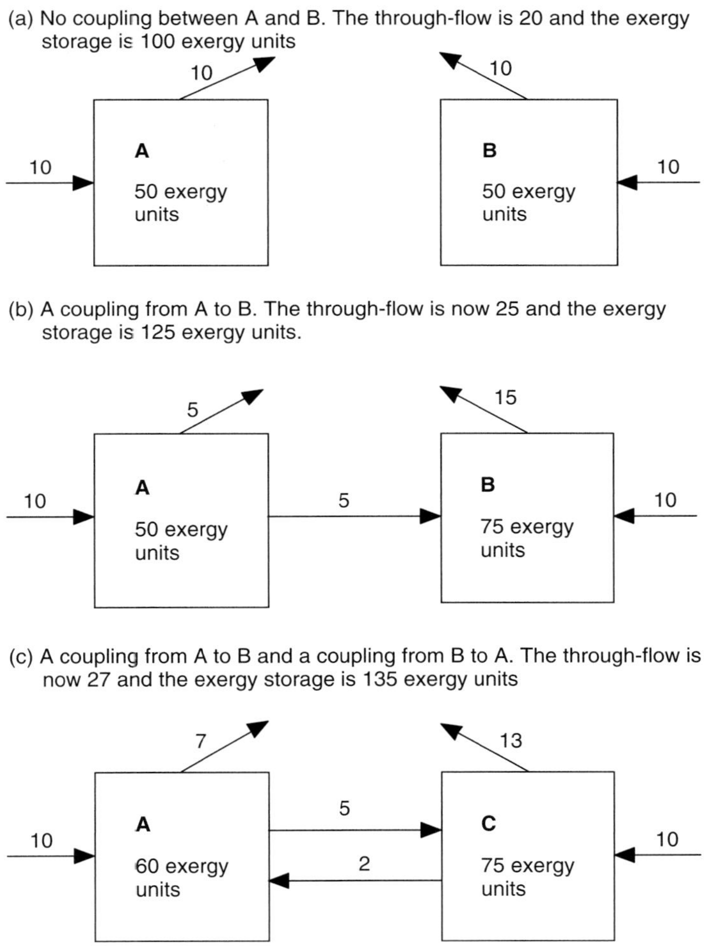

Figure 14.2 The figures a, b and c show an increasing number of connections that clearly imply increasing eco-exergy. Steady state is presumed. The sum of the outflows are all 20% of the eco-exergy in the compartments.

exited the system is not used by the second compartment, thereby increasing the total system through-flow, the exergy stored, and the average path length.

14.3 CALCULATIONS OF ENERGY AND ECO-EXERGY-BASED ASCENDENCY

The four networks in steady state that were used in Jørgensen et al. (2000) to illustrate the three growth forms (see also Jørgensen, 2002; Jørgensen and Svirezhev, 2004; Fath et al., 2004) have been applied here to calculate the eco-exergy storage and eco-exergy ascendency and compared with the previous energy-based ascendency. The below shown calculations are based on Jørgensen and Ulanowicz (submitted). Figure 14.3 shows a four-component network that has only cycling of the energy. The network is able to capture 5 units of energy, the flows are donor-regulated by a coefficient of 0.8. The sum of the outputs $1.5 + 1.5 + 1.1 + 0.9 = 5.0$ is due to the steady state conditions and the input $= 5.0$. They represent the energy used for maintenance, that is respiration. The network in the Figure 14.4 captures 10 units of energy, twice as much due to growth of the biomass, which determines how much solar radiation up to the physical maximum of about 75–80% the vegetation can capture. All the numbers in Figure 14.4 are twice as much as the numbers in Figure 14.3. Figure 14.5 represents the growth of the network. An energy transfer from component 3 back to component 1 is added, and the extra linkage in the network implies that the energy storage and ascendency are bigger in Figure 14.5 than in Figure 14.4. Figure 14.6 represents the growth in information and has a relatively smaller respiration. Both the eco-exergy stored in the network and the ascendency are bigger in Figure 14.6 than in Figure 14.5. The last three networks, Figures 14.4–14.6, have been repeated but now by using the eco-exergy storage and transfer. The eco-exergy is found as biomass $\times$ β, where $\beta = RTK$ is normalized to

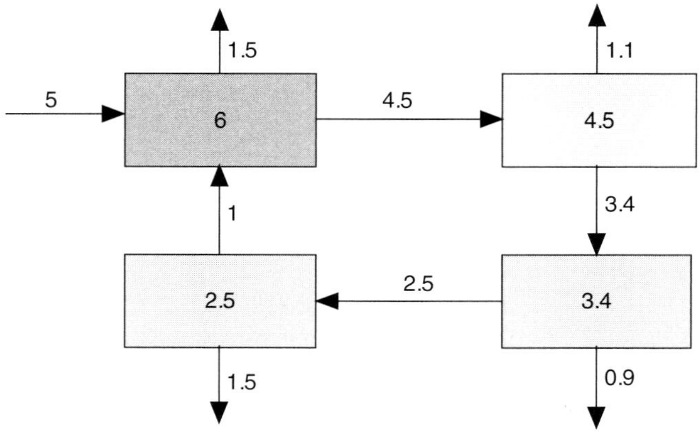

Figure 14.3 Reference network.

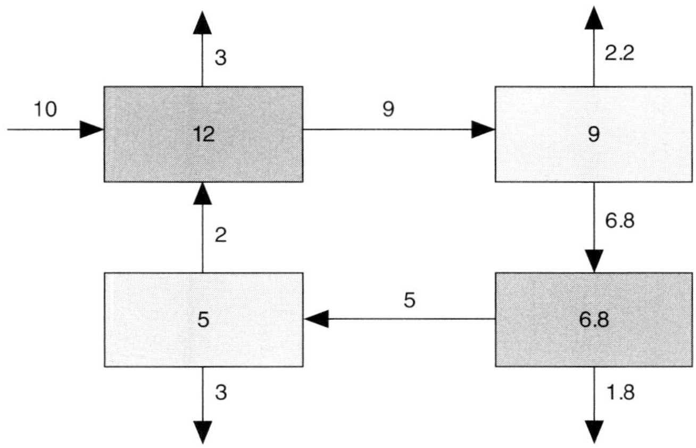

Figure 14.4 The input is twice the input in Figure 14.3.

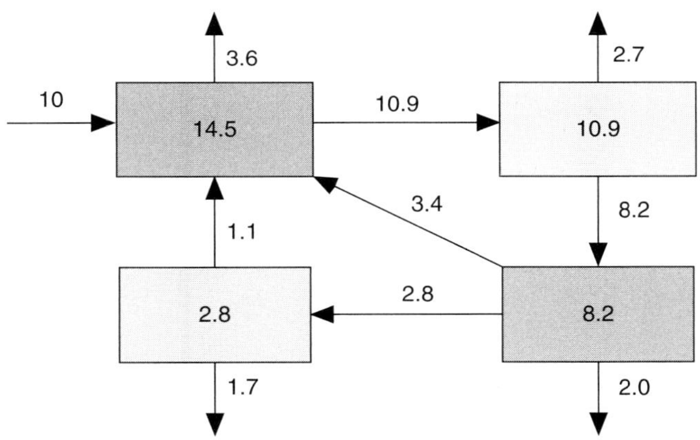

Figure 14.5 An extra flow is added from component 3 to 1, compared with Figure 14.4. The last figure, Figure 14.6, represents a growth in information corresponding to lower respiration rates.

express eco-exergy in detritus equivalents; $\beta = 1.0$ for detritus (dead organic matter). A list of values for different organisms have been published in Jørgensen et al. (2005).

Let us presume that all the networks in Figures 14.4–14.6 represent an aquatic ecosystem, where the first component is phytoplankton, which means that $\beta = 20$, the second component is zooplankton, which means that $\beta = 100$, the third component is fish, which means that $\beta = 500$, and the fourth component is detritus, which means that $\beta = 1.0$. Figure 14.4 will thereby be changed to Figure 14.7, Figure 14.5 to Figure 14.8 and Figure 14.6 to Figure 14.9.

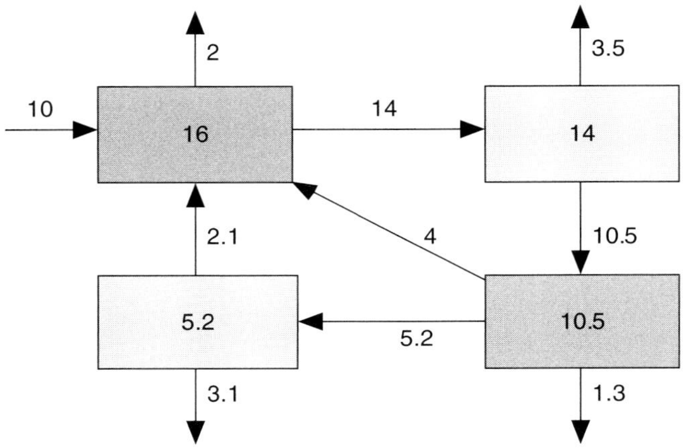

Figure 14.6 This network represents a growth in information corresponding to lower respiration rates.

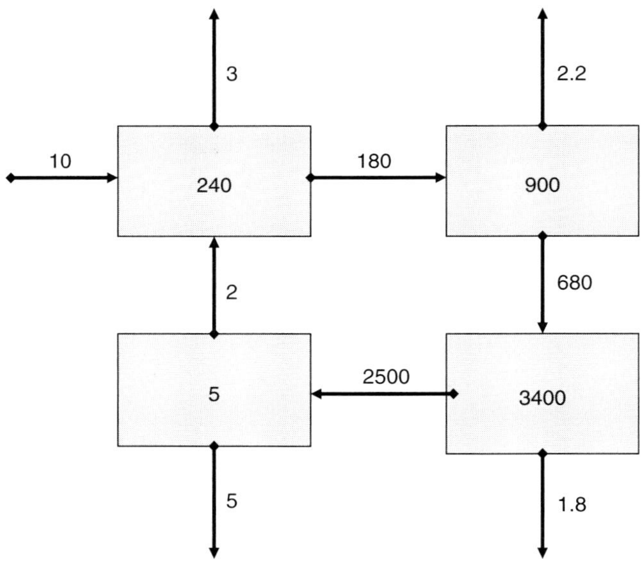

Figure 14.7 Same network as in Figure 14.4, but eco-exergy is calculated instead of energy.

The energy or eco-exergy storage, the energy or eco-exergy ascendency and the energy or eco-exergy power have been calculated based upon Figures 14.3–14.9. The results are shown in Table 14.1.

The results show not surprisingly that it is necessary to calculate the eco-exergy-based ascendency, if it has to be compared with the eco-exergy storage. The eco-exergy

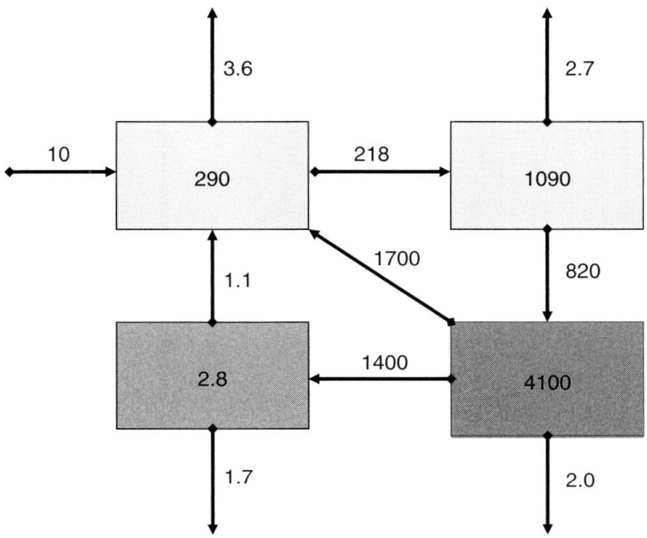

Figure 14.8 Same network as in Figure 14.5, but eco-exergy is calculated instead of energy.

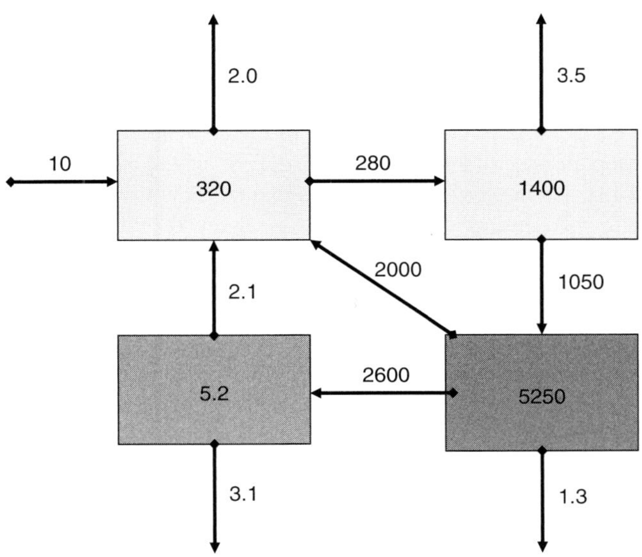

Figure 14.9 Same network as in Figure 14.6, but eco-exergy is calculated instead of energy.

Table 14.1 Results and comparisons figures 14.4–14.10

Figure	Energy or eco-exergy storage	Energy or eco-exergy ascendency[a]	Energy or eco-exergy power[a]
14.4	16.4	34.65	11.4
14.5	32.8	69.29	22.8
14.6	36.4	65.10	26.4
14.7	45.7	86.92	35.8
14.8	4545	3553	3362
14.9	5483	4168	4139
14.10	6975	4666	5932

[a] If the eco-exergy unit is kJ/m^2, the unit for ascendency and power could be $kJ/24\,h\,m^2$.

ascendency should therefore be added to the eco-exergy storage to express what is obtained by changing the network. The sum could be applied as a goal function in structurally dynamic modelling to account for the contribution from network changes. The increases in percentage for different changes have been calculated (see Table 14.2). The increase from Figure 14.3 to 14.4 is not included in the table because it is just a doubling of all flows, exergy storage, ascendency and power.

The eco-exergy storage increases significantly more than the energy storage 20.8% versus 9.1% by adding an extra transfer from component 3 to 1. The reason is the increased cycling is utilized particularly by the later components in the food chain and they have a higher β-value. This is completely in accordance with the network rules published by Jørgensen and Fath (2006). The eco-exergy storage increases 25.5% by adding more information and thereby reduce the respiration from Figure 14.8 to 14.9. It is the same increase as for energy—Figure 14.5 to 14.6. The less energy used for respiration is utilized by the very component. It means that more energy or eco-exergy is available for the flows, which explains that the power increases by this change of the network (Figure 14.5 to 14.6 and Figure 14.8 to 14.9) more than the energy or the eco-exergy stored.

There is a drop in ascendency from the network in Figure 14.4 (69.29) to that in Figure 14.5 (65.13). The reason for the drop is that the new flow from component 3 to 1 adds ambiguity to the network. Notice that ascendency is NOT simply proportional to the total system throughput. If all possible flows are realized in the network, the

Table 14.2 Increases in energy or eco-exergy storage, ascendency and power

Figures	Storage (%)	Ascendency (%)	Power (%)
From 14.5 to 14.6	9.1	–6.0	15.5
From 14.6 to 14.7	25.5	33.5	32.2
From 14.8 to 14.9	20.8	17.3	23.1
From 14.9 to 14.10	25.5	12.0	43.3
From 14.5 to 14.7	39.6	27.1	57.5
From 14.8 to 14.10	53.4	31.3	46.6

ascendency is zero, because ambiguity about where a quantum in any compartment will flow next is maximal. It doesn't matter how strong the flows are, the ascendency will remain zero, so long as all the magnitudes remain equal.

The very idea behind ascendency was to modify the total system throughput to quantify how organized (definitive) flow in the system is. The three networks, 14.4–14.6, all have the same total throughput (namely 10), but their ascendencies are decidedly different and reflect the degree of unambiguity in each one.

Now, looking back at Figures 14.4 and 14.5, we notice that flow in Figure 14.4 is not very ambiguous. In particular, if a quantum is in compartment 3, the only other compartment to which it can flow is to compartment 4. (Of course, it could also leave the system as export.) In Figure 14.5, by contrast, if a quantum is in compartment 3, there is some uncertainty as to whether it will flow to compartment 4 or to compartment 1. This lowers the ascendency, even though there is more total flow in Figure 14.5 than in Figure 14.4.

Notice how each flow generates one and only one term in the formula for ascendency. In particular, the contribution of T_{31i} in Figure 14.5 to the ascendency is

$$A_{31} = T_{31} \log\left(\frac{T_{31}T_{..}}{T_{3.}T_{.1}}\right),$$

or

$$A_{31} = 3.4 \log\left(\frac{3.4.46.4}{8.2.14.5}\right) = 3.4 \log(1.327) = 1.388$$

That is, T_{31} contributes proportionately less than its magnitude to the ascendency.

T_{34} (= 2.8) contributes 7.001 to the ascendency. So the total of T_{31} and T_{34} in Figure 14.5 is 8.389. Contrast this to the amount that T_{34} in Figure 14.4 contributes (= 13.27), and the shortfall in Figure 14.5 becomes apparent.

Calculations of eco-exergy storage, ascendency and power show that they are following the same trends when changes are made except for the energy-based ascendency, when an extra connection is added. While the storage and the power are increased, ascendency is decreased by the addition of an extra net flow due to addition of ambiguity.

The eco-exergy-based ascendencies and powers are significantly higher than the energy-based ones. If the full consequences of changing a network should be calculated, for instance in a structurally dynamic model or by illustrating the evolutionary increase of ecological networks, it would therefore probably be advantageous to use eco-exergy ascendency. The experience of using eco-exergy as goal function in structurally dynamic modelling has been positive in totally 18 case studies. All the case studies have, however, been on changes in the properties of the key species. It cannot be excluded that major changes of the network could be covered more correctly by using the sum of eco-exergy and the eco-exergy-based ascendency as the goal function. To illustrate the evolution of ecological networks, both energy and eco-exergy-based ascendency will be used in the next chapter.

15

The evolution of networks

15.1 THE HORIZONTAL EVOLUTION

Both the growth and development of networks and of the information level are dependent on the species that are available for the ecosystems to move as far as possible away from thermodynamic equilibrium (Vou Bloh et al., 2003). The evolution has, as already underlined, been both vertical—that is, the eco-exergy per gram biomass has increased – and horizontal, that is the number of species has increased. The increase of the eco-exergy per gram biomass during the evolution, from the prokaryote cells about 3800 million years ago to *Homo sapiens* today, has been presented in the Second Movement and in Jørgensen (2007b). It would therefore be interesting to attempt to describe in a similar manner the result of the horizontal evolution. Which development of the possible ecological networks can we expect as a result of the horizontal evolution and which increase in eco-exergy will this development of the networks imply?

We do not know all the species from fossil records, because it is probably that only a small fraction of the previous species has been found as fossils. The evolution of the biodiversity has therefore, in Chapter 13, been presented by the number of families. On the other hand, the most dominant species are probably represented by the fossils. It may therefore with approximations be possible to set up typical and most probable ecological networks that have evolved due to the horizontal evolution, because we have a certain although limited knowledge of the dominant species, their properties and food items. As we do not have knowledge to the ecological networks as result of the evolution, we have to make qualified guesses, presenting possible and probable ecological network. From calculations at different evolutionary steps of

1. the power = the sum of energy flows
2. the eco-exergy power = the sum of eco-exergy flows
3. the ascendency, calculated based on energy flows
4. the ascendency, calculated based on eco-exergy flows,

we could obtain a first rough and approximate estimation of the evolution of ecological network. These calculations can directly be compared with the result of the vertical evolution obtained in the Second Movement. We may thereby be able to draw a picture of how more and more complex ecological networks have evolved from a wider and wider spectrum of more and more advanced species.

15.2 PRESENTATION OF A POSSIBLE AND PROBABLE EVOLUTION OF NETWORKS

The evolution of possible and probable networks, based on what may be denoted qualified guesses, is presented in Figures 15.1–15.22. Notice all figures with odd numbers are based on energy flows and all figures with even numbers are based on eco-exergy flows. All the diagrams are based on an energy input of 10 energy units.

It is presumed that the ecological networks capture the same amount of energy from the solar radiation, because the idea is to show the differences in the structure of the networks—not in their ability to capture solar radiation, which is discussed in the Coda.

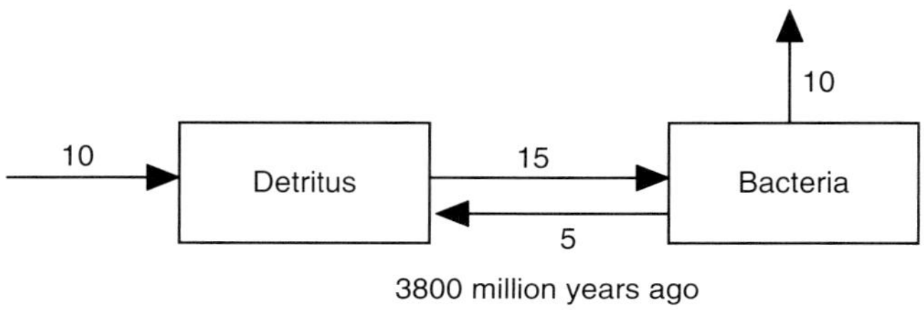

Figure 15.1 The Earth was inhabited only by primitive cells 3800 million years ago, which were living on detritus.

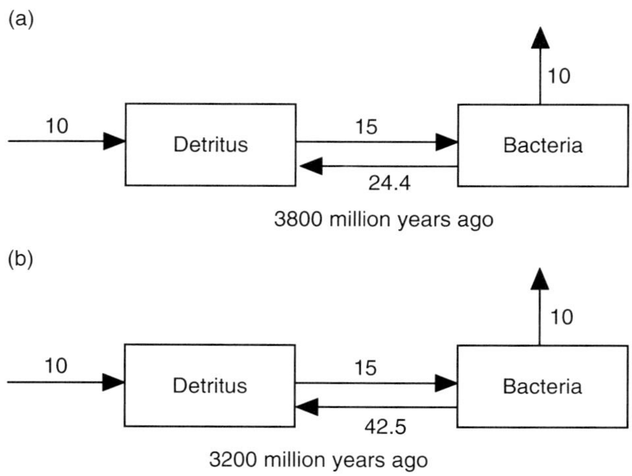

Figure 15.2 (a) Energy is replaced by eco-exergy in Figure 15.1. (b) 3200 million years ago, the energy flows were the same as in Figure 15.1, but the primitive cells were replaced by prokaryote cells.

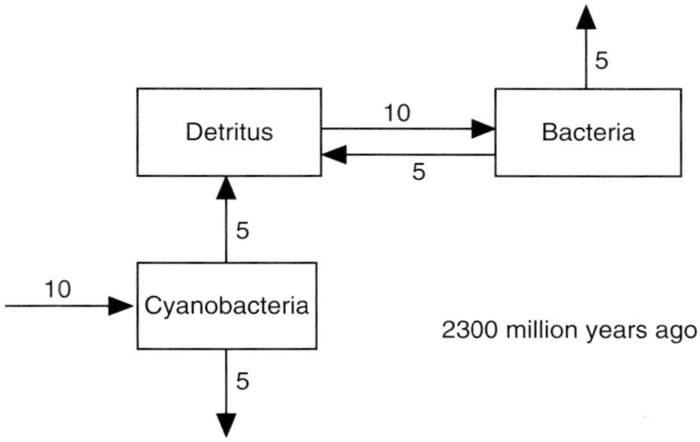

Figure 15.3 Photosynthesis because possible by cyanobacteria 2300 million years ago, which gave rise to a more complex network. Cyanobacteria are also prokaryote cells.

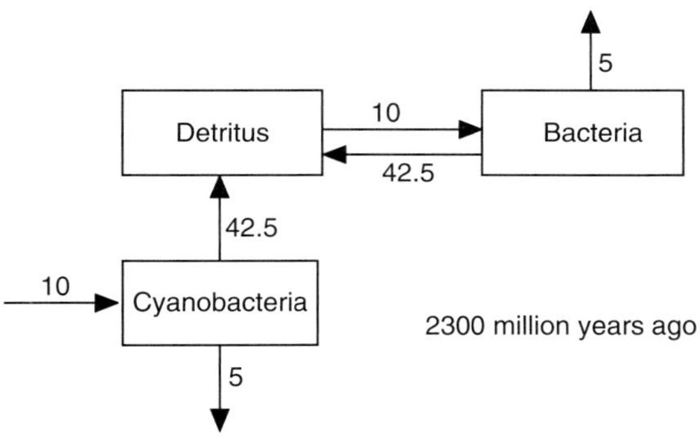

Figure 15.4 Energy is replaced by eco-exergy in Figure 15.3.

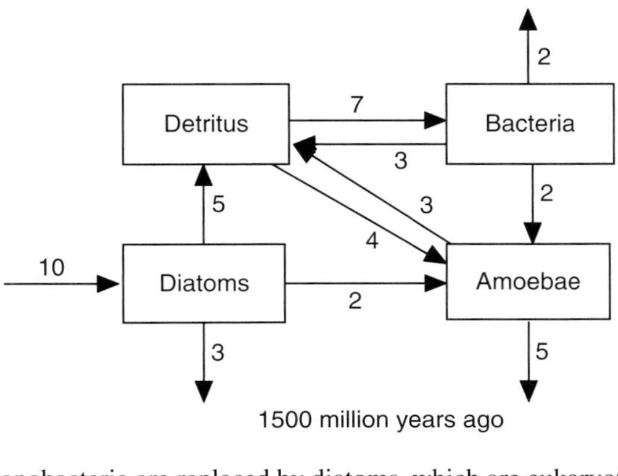

Figure 15.5 The cyanobacteria are replaced by diatoms, which are eukaryote cells. The amoebae has been introduced and the ecological network consists now of four interacting components.

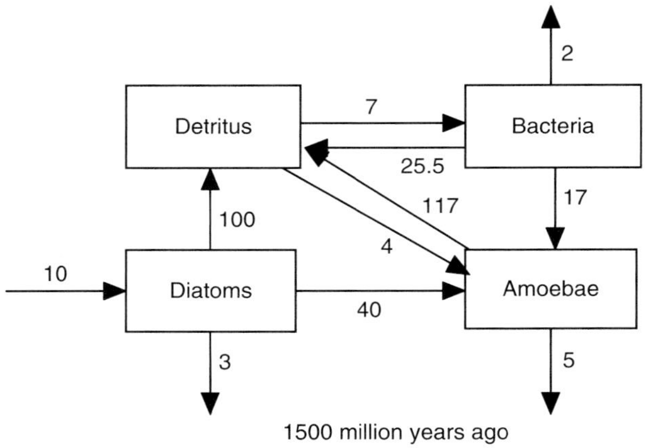

Figure 15.6 Eco-exergy has replaced energy in Figure 15.5.

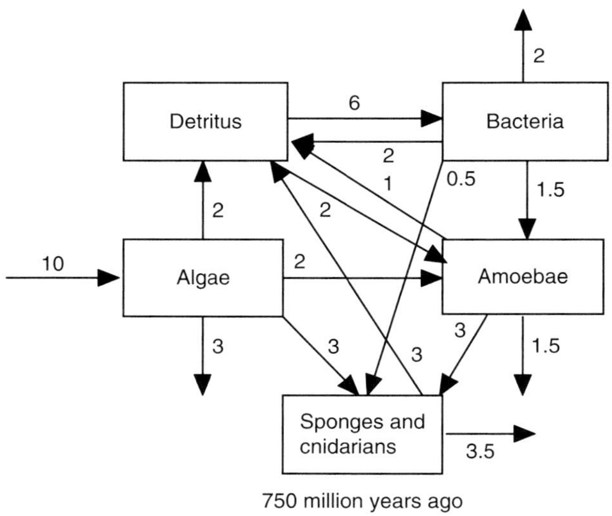

Figure 15.7 The first polycellular organisms emerged before the Cambrian Explosion.

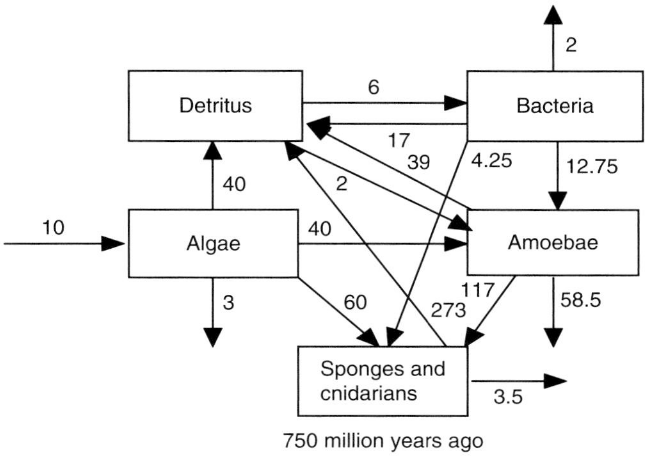

Figure 15.8 Energy is replaced by eco-exergy in Figure 15.7.

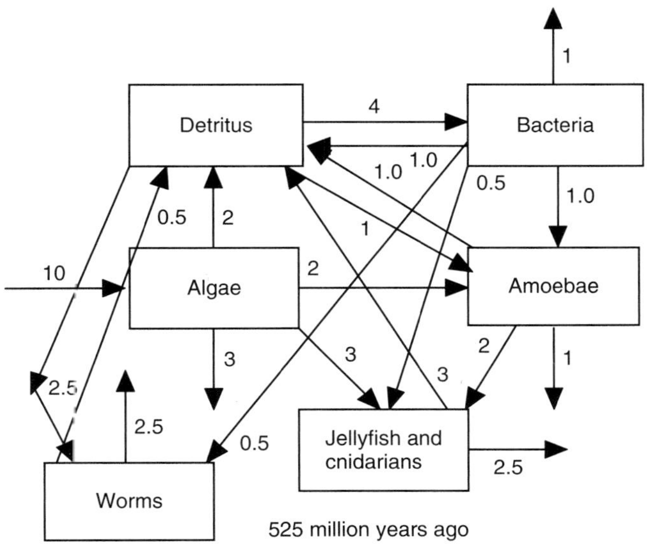

Figure 15.9 More polycellular organisms emerged as a result of the Cambrian Explosion.

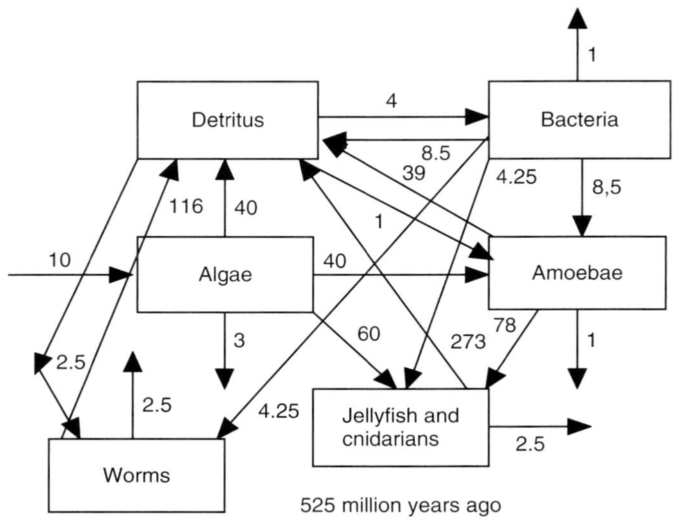

Figure 15.10 Eco-exergy has replaced energy in Figure 15.9.

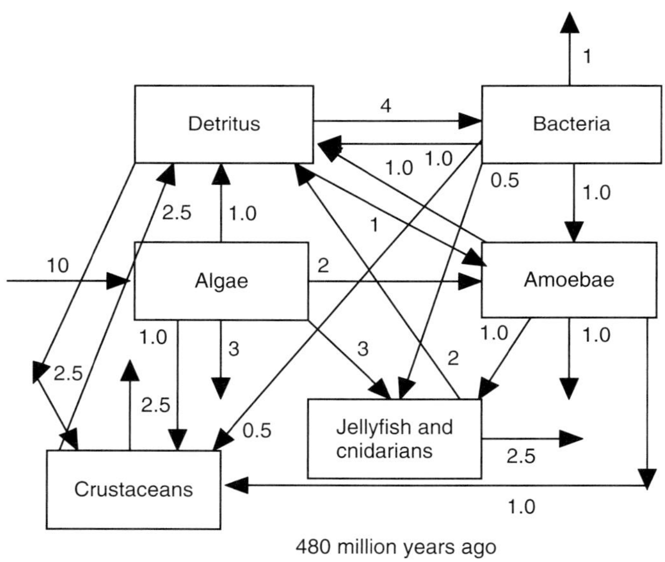

Figure 15.11 The result of the Cambrian Explosion is more advanced organisms.

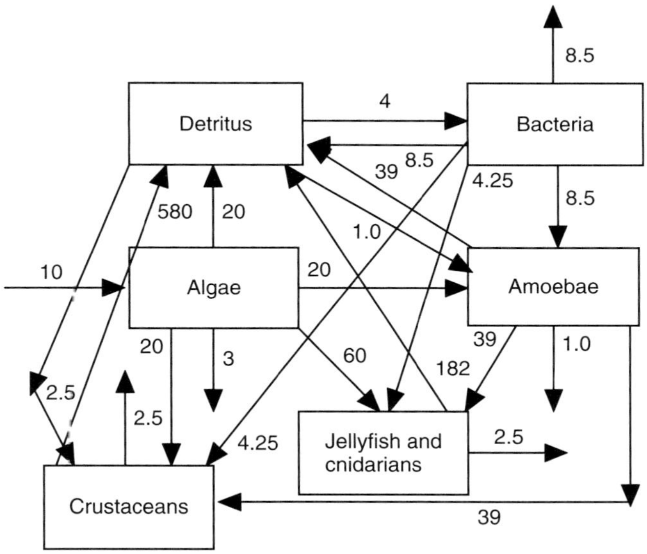

Figure 15.12 Eco-exergy has replaced energy in Figure 15.11. A wide spectrum of crustaceans have merged as a result of the Cambrian Explosion. The crustaceans have significantly higher β-values than the other organisms in the diagram, which of course implies that the eco-exergy flows are increased considerably compared with the previous diagrams.

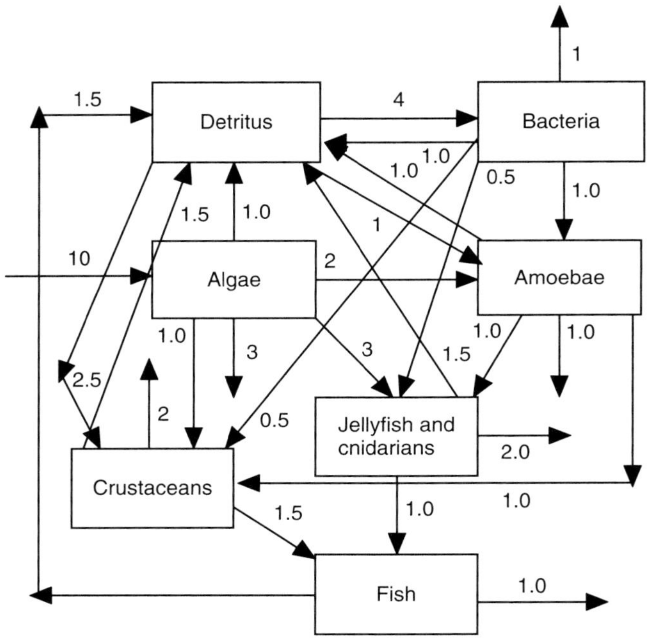

Figure 15.13 The first primitive fish species have emerged. They are the first vertebrates and have considerably higher β-values. The biodiversity has increased significantly since the Cambrian Explosion started 85 million years earlier (see also Table 19.1 in Section 9.1).

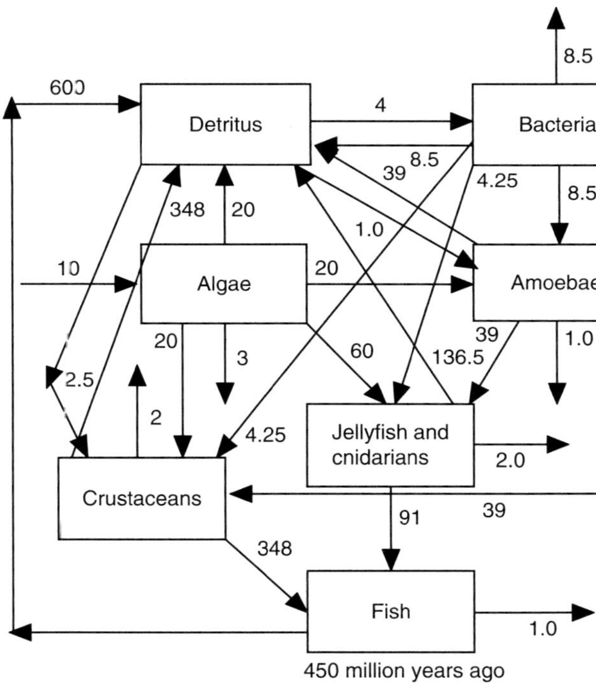

Figure 15.14 Similar to Figure 15.13, but energy is replaced by eco-exergy. A β-value of 400 is used for the first primitive fish, while the β-value for fish previously has been indicated as 499.

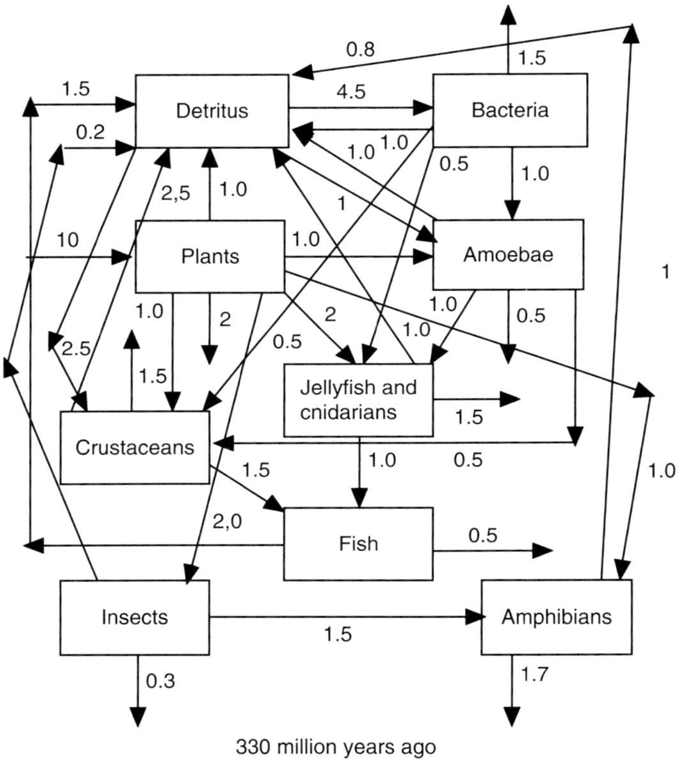

Figure 15.15 Amphibians, insects and terrestrial plants have emerged. All three contributed to a significant increase in the ecological network and the energy flows, the power.

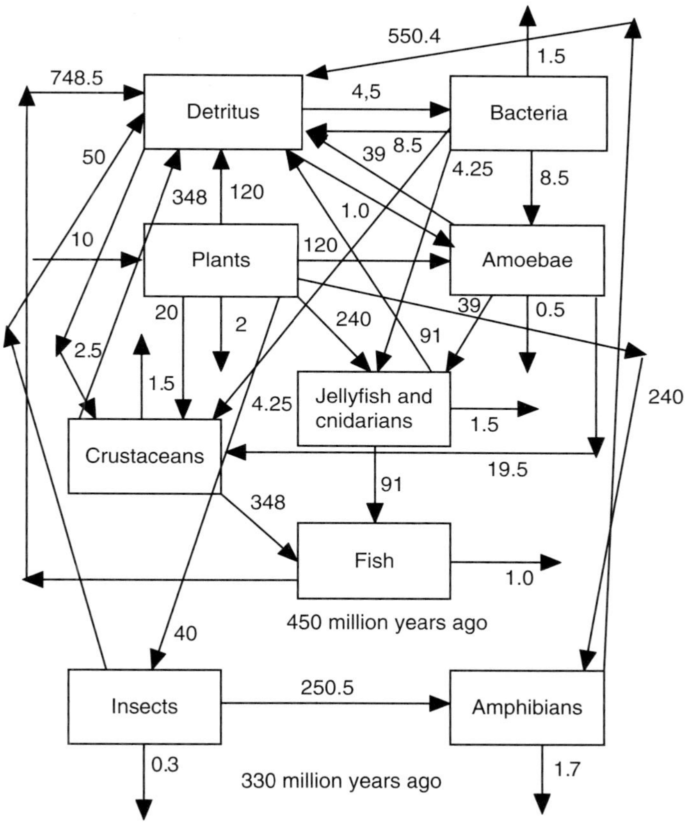

Figure 15.16 The relatively high β-values of amphibians, insects and terrestrial plants imply that the eco-exergy flows and the eco-exergy ascendency are considerably higher than the energy flows and the energy ascendency. A β-value of 499 is now used for the more advanced fish species, which evolved 330 million years ago.

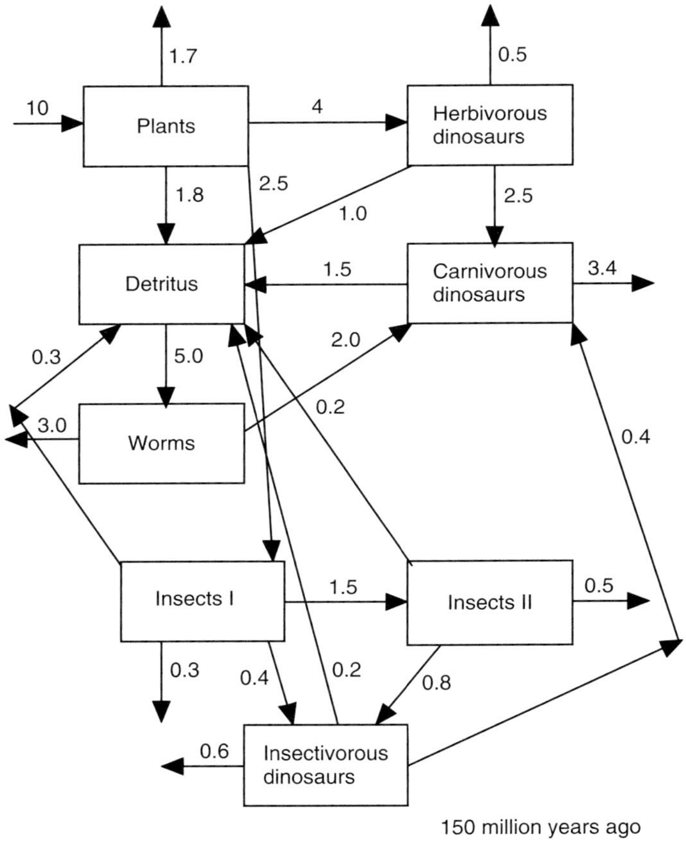

Figure 15.17 A wide spectrum of dinosaurs placed in different knots of the ecological networks and with different food sources have emerged.

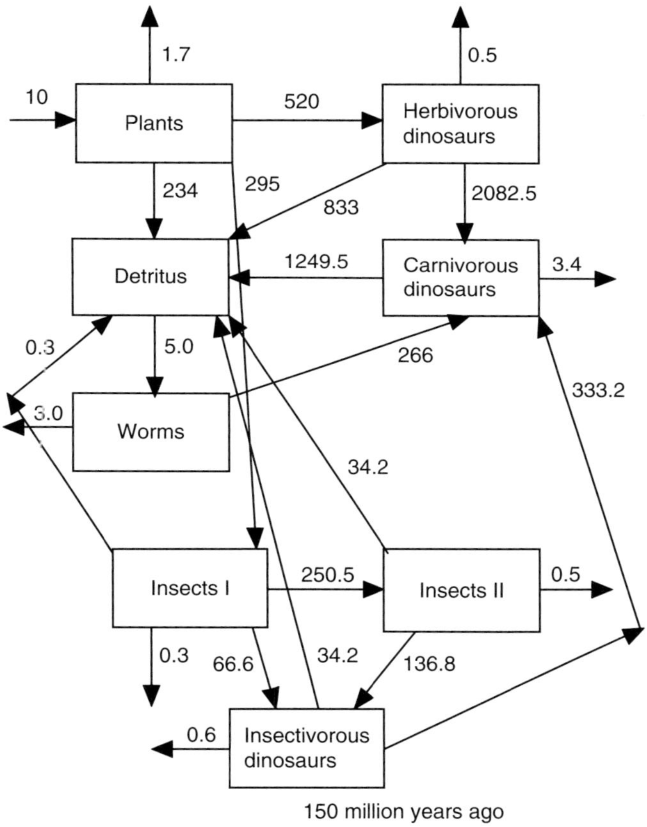

Figure 15.18 Eco-exergy has replaced energy in Figure 15.17. Due to the high β-values of dinosaurs the eco-exergy flows and the eco-exergy ascendency are significantly higher than the energy flows and the energy ascendency, which is in accordance with the results in Chapter 14.

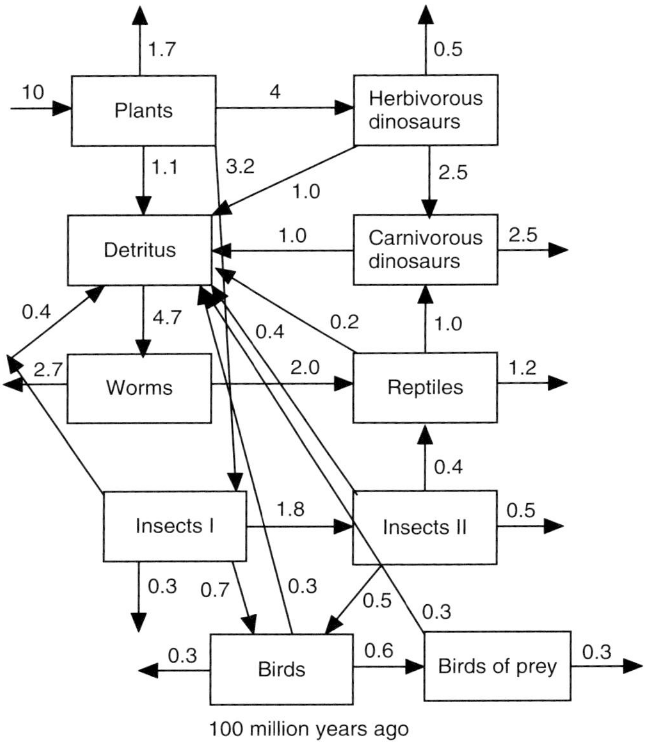

Figure 15.19 Birds, most probably evolved from dinosaurs, have been introduced. A relative complex ecological network is the result.

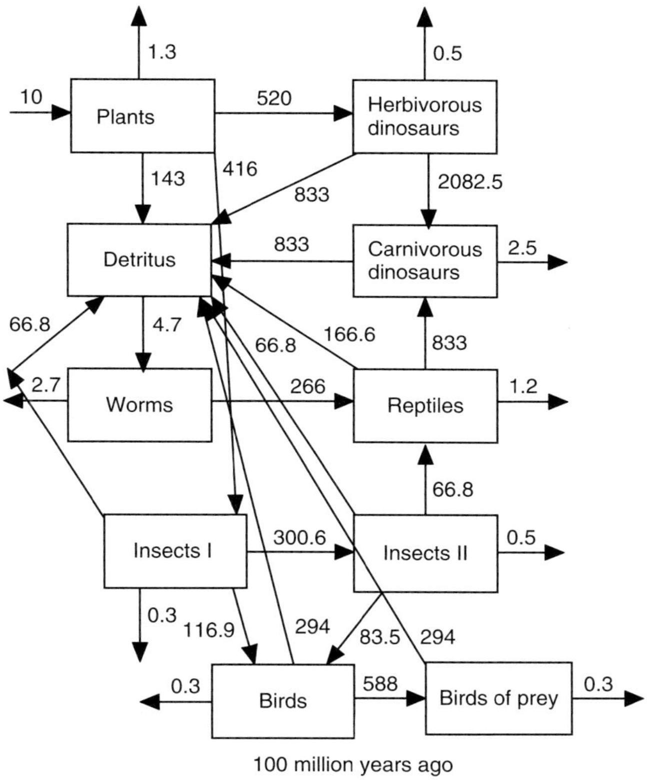

Figure 15.20 The high β-value of birds, 980, contribute to high eco-exergy flows and a high eco-exergy ascendency.

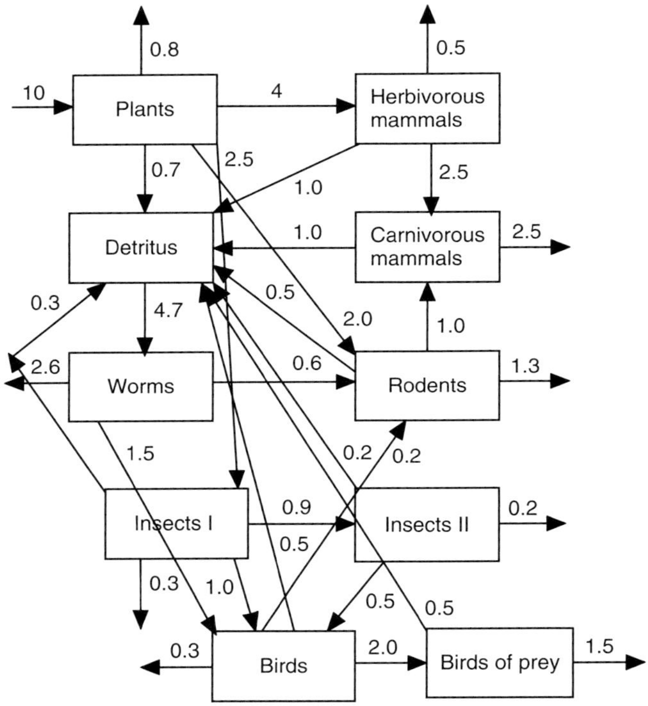

Figure 15.21 Three types of mammals are included in the ecological network shown: herbivorous mammals, carnivorous mammals and rodents. The ecological network is probably not more complex than the ones in Figures 15.17 and 15.19. The evolution from the dinosaurs to the mammals has, to a greater extent, been vertical. The same is probably the case for the evolution from the wide spectrum of mammals to man.

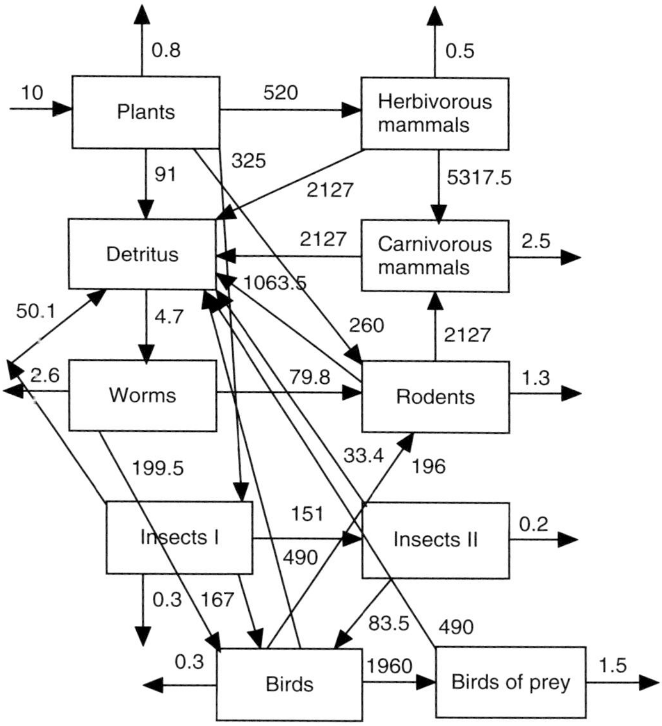

Figure 15.22 The high β-value of mammals implies that the eco-exergy flows and the eco-exergy ascendency are considerably higher than energy flows and the energy ascendency, or expressed differently, the evolution from 150 million years ago to 35 million years ago has, to a greater extent. been vertical. It is probably also the case for the evolution from general mammals to man, that is the evolution of the last 35 million years.

15.3 THE THERMODYNAMIC INTERPRETATION OF THE HORIZONTAL EVOLUTION

The results of calculating energy power, eco-exergy power, energy-based ascendency and eco-exergy-based ascendency for the networks shown in the 22 figures, 15.1–15.22, are summarized in Table 15.1.

The energy ascendency approximately doubled during the evolution from the primitive prokaryote to the mammals. It may therefore be possible to conclude that the evolution has had a tendency to increase slightly the organization of the networks expressed by the ascendency and the ambiguity about where a quantum in any compartment will flow next has decreased correspondingly. The energy power is roughly not increased or increased very little. The eco-exergy power and ascendency have increased 300–400 times, which can be explained by the increase of eco-exergy of the components (see Second Movement). The eco-exergy per gram of biomass has increased in the same period from 91 to 39,775 kJ/g or 437 times. The increase of biodiversity as we have presented in Chapter 13 can therefore not have caused an increase of the eco-exergy ascendency, but the increase seems entirely to be due to the increase in eco-exergy per gram of biomass. These results are consistent with the results presented in Chapter 14, where it was shown that ascendency is not necessarily increased by adding more flows to the network.

The networks have, however, increased in complexity and the same amount of energy—the 10 units—can support more and more biomass and eco-exergy, because the growth of the network implies that the mass, energy and eco-exergy due to cycling can be utilized better and better. Table 15.2 has found the amount of biomass expressed as kJ (1 g has 18.7 kJ of energy) that the network can sustain, presuming that the network is a donor-based network where the outflows to the other components all are 20% of donor's energy content. Similarly, the eco-exergy sustained by the various networks has been found. The energy content of the various components has been multiplied by the β-values (see Table 1.1). The results of these calculations are shown in Table 15.2.

Table 15.1 The results of the networks in Figures 15.1–15.22

10^6 years ago	Energy power	Eco-exergy power	Energy ascendency	Eco-exergy ascendency
3800	40.0	59.4	48.7	52.7
3200	40.0	77.5	48.7	57.7
2300	40.0	115	60.0	96.7
1500	46.0	331	62.2	161.5
750	46.0	631	65.4	495
525	44.0	699	65.1	519
480	45.0	1,052	62.9	675
450	47.5	1,911	69.3	1,555
330	52.5	3,409	86.0	3,510
150	44.1	6,361	86.2	7,606
100	46.1	7,995	97.6	12,174
35	48.1	17,883	102	20,845

Table 15.2 Biomass and eco-exergy sustained by the networks in Figures 15.1–15.22

10^6 years ago	Biomass sustained by the networks	Eco-exergy sustained by the networks
3800	100	220
3200	100	288
2300	100	763
1500	130	2,045
750	120	3,050
525	120	3,092
480	125	5,218
450	132	9,677
330	135	16,412
150	121	38,430
100	137	44,090
35	155	84,737

The results show that the network 35 million years ago could sustain 55% more biomass, which must be due entirely to a more effective use of the incoming energy by the network. The eco-exergy of the networks has increased about 400 times or about the same as the eco-exergy per gram. However, the network consists of many different organisms with different eco-exergy per gram. The development of the network has, in other words, provided the entire network with an increase in the information-based eco-exergy that corresponds to the most developed organisms. The development of the networks has made it possible for the entire ecosystem to increase its eco-exergy content or work capacity to the same level—about 400 times during the evolution—as the most developed organisms, that is the organisms with most eco-exergy. The evolution of the ecological networks has been possible due to the horizontal evolution—the evolution of the species richness.

16

Summary of the evolution of eco-exergy and discussion of the evolutionary possibilities in the future

16.1 OVERVIEW OF THE VERTICAL EVOLUTION

In the Second Movement, the evolutionary steps have been briefly mentioned and the obtained eco-exergy density in kJ/g and eco-exergy flow rate in J/kg s have been found for each important evolutionary step to be able to express the evolution thermodynamically. The eco-exergy flow rates for the steps of the physical evolution, galaxies $\rightarrow$ stars $\rightarrow$ planets, before the evolution of life, have been calculated in Chapter 5. Figure 16.1 shows a graph of the eco-exergy density as a function of time and Figure 16.2 illustrates the eco-exergy flow rate as a function of time. Both functions show a more and more rapid increase. In this context, it is important to remember that eco-exergy expresses biomass $\times$ information and that it is the information factor that is increasing more and more rapidly. As previously discussed, the growth of biomass is limited in accordance with the amount of the limiting elements, while the information still has the possibility to increase orders of magnitude (see Chapter 2). The evolution has therefore been towards a more and more effective biochemical control, a more and more effective use of energy and more and more sophisticated life forms, which is expressed thermodynamically in the two plots in Figures 16.1 and 16.2.

If semi-logarithmic plots are made of these two figures, the graphs in Figures 16.3 and 16.4 are produced. The Cambrian Explosion and the increase by the emergence of the first life are seen clearly on these semi-logarithmic graphs. As seen in Figure 16.3 the eco-exergy density has increased close to exponentially, although the increase has been slightly faster than exponentially when life emerged and since the Cambrian Explosion. Exponential growth means that eco-exergy density grows in accordance with a first-order reaction:

$$(\mathrm{dex}/\mathrm{vol}) = r\,\mathrm{d}t$$

where ex is the eco-exergy, vol, the volume, and r, a rate constant. Figure 16.3 indicates that the slope is more than r at the emergence of the first primitive cell and after the

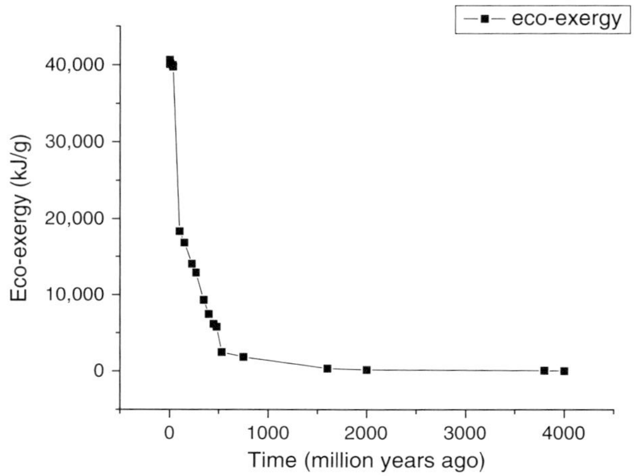

Figure 16.1 Eco-exergy density in kJ/g is plotted versus time. Notice the abrupt increase at the beginning of the Cambrian time, about 540 million years ago. Also the enhanced increase with the mammals, hominoids and humans can be seen clearly on the graph.

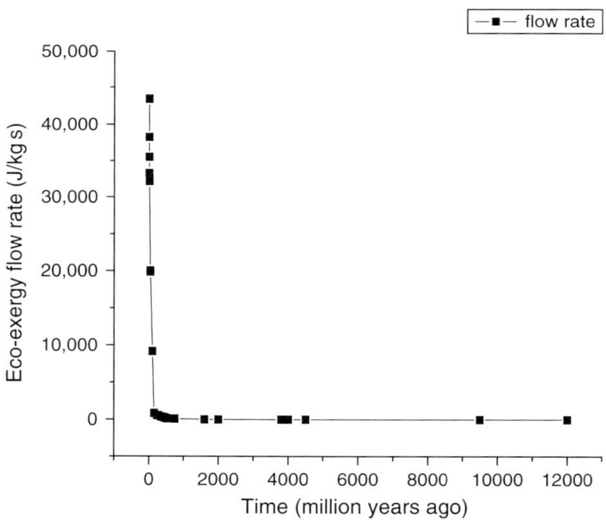

Figure 16.2 Eco-exergy flow rate versus time is shown.

Cambrian Explosion. Particularly in the last few million years, the slope is clearly more than the average slope, represented by the regression line.

The eco-exergy flow rate has increased faster than a first-order reaction, with a clear bigger slope after the Cambrian explosion and when life emerged (see Figure 16.4). In the last few million years the slope has increased even further (see also Jørgensen, 2007b).

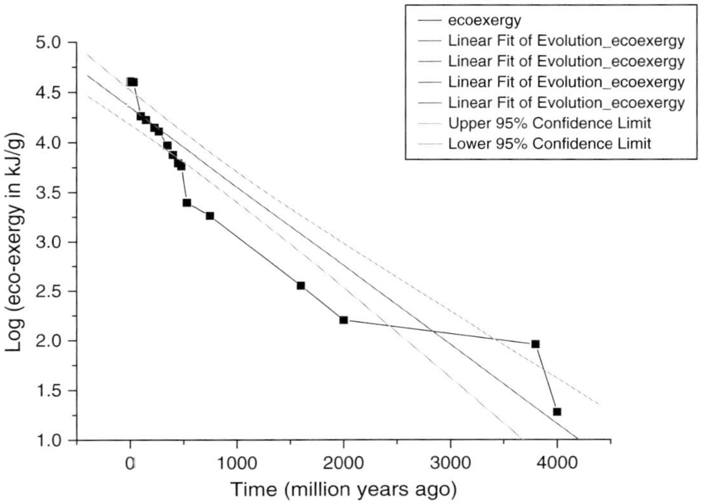

Figure 16.3 Semi-logarithmic plot of eco-exergy density versus time. The plot is with a rough approximation linear, which implies that the eco-exergy density has increased exponentially. The correlation coefficient is 0.95. The confidential intervals plus or minus one standard deviation are indicated. The increase has, however, been particularly fast since the Cambrian Explosion and just after the emergence of the first primitive life.

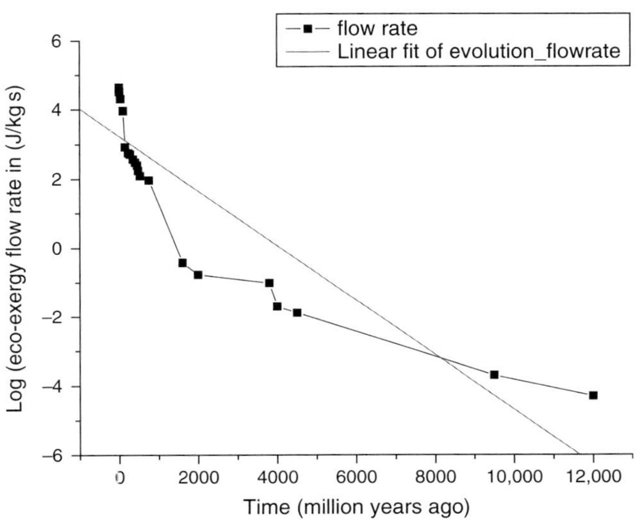

Figure 16.4 A semi-logarithmic plot of the eco-exergy flow rate versus time is shown. The increase has clearly been faster than exponentially.

Both the eco-exergy density and the eco-exergy flow rate have increased significantly since the mammals have conquered the Earth 65 million years ago. As previously mentioned it cannot be excluded that the values calculated for the dinosaurs are under-estimated, because we have considered them as non-homeothermic in our calculations. An interesting question in this context as previously mentioned: why did the mammals not take over previously, as they emerged already about 200 million years ago? A tentative answer could, as previously discussed, be that the dinosaurs were already abundant and it is easier to fill out an empty ecological niche than to replace already well-functioning organisms in an occupied niche. It is similar to the frequently made observations in ecological models, where the initial concentration value of an organism is crucial for its survival.

The eco-exergy flow rate expresses, after the emergence of life, mainly the rate of exchange of information. More and more information has been stored in the biomass but the rate of information transfer has increased rapidly due not only to the amount of information that is transferred per information packet or message but also to the turn overrate—a factor 1000 from the prokaryotes to the humans (see the calculations in the Second Movement).

As the energy consumption and the transfer rate are particularly high in the brain, the rapid increase during the past about 1 million years expresses and includes the increased level of communication, with the main factor being language, the use of which demands a well-coordinated brain function. Communication today has been enhanced several times, but it is no longer due to increased storage of information in the genes, but due to cultural development.

The rapid growth of the eco-exergy stored and the exergy flow rate when the first primitive cells emerged on Earth (see Figures 16.3 and 16.4) indicate that the first primitive cells may not have been developed on the Earth, but that they may have been brought to us from the open space, for instance, by meteorites. The evolution is going faster and faster and the increase of complexity is exponential or maybe even faster than exponential. If it is assumed that the eco-exergy stored has increased exponentially as stated in Moore's law (complexity of certain systems increases exponentially), an extrapolation to a complexity corresponding to no life (let us use log eco-exergy = 0.5 corresponding to the free energy of many low molecular organic compounds; see Figure 16.3) will tell us that the life may have emerged about 5–6 billion years ago, not 3.8 billion years ago. Sharov (2006) has used a plot of the functional non-redundant genome size versus time and presumed exponential increase of the complexity to conclude that the origin of life should be dated 10 billion years ago. In this context it is not important whether the origin of life is 5, 6 or 10 billion years ago, but it is important to conclude that the origin of life is older than the Earth (about 4.5 billion years).

The Earth had probably the right temperature for life about 4 billion years ago. It *is* therefore surprising that already 200 million years later, the first living cells emerged. The first steps of the evolution have inevitably been slow, because the evolution of life must build on what has already been achieved, and 4 billion years ago it was very little, probably only the formation of some polypeptides at the most. The first steps had no workable solutions to build on, and it is expected that the first primitive cells have evolved from numerous trial-and-error experiments. The relatively fast emergence of life after the right temperature has been obtained seems therefore, as the above-mentioned extrapolations of Figure 16.3 and of Sharov's plot, the functional non-redundant genome size versus time, to point towards an origin of the first life outside the Earth.

16.2 THE INFORMATION SOCIETY

The information in a book having 10^6 signs can be calculated from Boltzmann's equation:

$$\text{Energy content of the information/sign} = kT \ln W \tag{16.1}$$

where k is Boltzmann's constant $= 1.38 \times 10^{-23}$, T is the absolute temperature and W is the number of microstates that yields one particular macrostate. If we presume that there are 55 possible signs, the total energy content of the information in a book can be found as

$$\text{Eco-exergy in a book} = 1.38 \times 10^{-23} \times 300 \times 10^6 \times \ln 55^{1\,000\,000} = 16.56 \times 10^{-9}\ \text{J}$$

About 10^{10} books, journal's and newspapers per year imply a total eco-exergy per year for written material on about 200 J, of course the eco-exergy of the paper, which is wasted as heat by incineration.

The films produced per year contain about 10 times as much information, the private photos about 100 times as much information, the TV transmissions 1000 times as much information and the private telephone conservations and faxes 10,000 times as much information as the books and newspapers, according to Jensen (2004). It means that books, newspapers, TV-transmissions, photos, telephone conversations and faxes together will contain information as eco-exergy corresponding to 2.2×10^6 J/yr, which is negligible compared with the information content of living organisms (see also the calculations in Jørgensen, 2006). The difference is that living organisms have information in many molecules (Avogadro's number is 6.2×10^{23}), so the enormous number of molecules carrying the information is explaining why the information stored in the society is minor compared with the information stored in the living organisms. It means that the β-value of books, films, photos, videos and so on is close to 1.00, meaning that the energy content of the hardware is much bigger than the exergy content of the "software". This is in contrast to the "software" content of organisms, where, for instance, mammals have more than 2000 times "software" eco-exergy than hardware eco-exergy, which is only the about 18.7 kJ/g. A new technological development—nano-technology—attempts to use the enormous information storing capacity per unit of volume on the molecular level. At present the information society can, however, still not compete with the biology on the information content.

16.3 THE FUTURE

In principle, we cannot increase the biomass very much due to the limiting elements (Liebig's Minimum Law) and thereby gain eco-exergy; but it is clear that we loose a significant amount of eco-exergy by deforestation and gain a lot by plantation of forest.

If, for instance, 1 ha of forest with 200 kg of biomass per m^2 (plants, trees, deers, insects, birds, etc.) with an average β-value of 180 (compare with Table 1.1, Chapter 1) is cleared, the loss of eco-exergy will be

$$\text{Eco-exergy loss} = 10{,}000 \times 200{,}000 \times 180 \times 18.7\,\text{kJ} = 0.7 \times 10^7\,\text{GJ/ha}$$

If we consider that the annual increase in eco-exergy of forest land is about, let us say, 2%, the loss in annual eco-exergy production would be 1.4×10^5 GJ/ha yr. The 1 ha cleared forest land may of course often be used as agricultural land in which case there will an annual production of about 20 *t* biomass with, let us use, an average β-value of 150 or 5.6×10^4/ha yr or 2.5 times less than for the forest. So, the first year after the deforestation, the eco-exergy capital loss is 0.7×10^7 GJ, followed by an annual loss of about 0.8×10^5 GJ/ha.

The global annual deforestation is difficult to estimate because deforestation also takes place illegally; but we could use the information given by FAO (2005). The total annual loss of forest cover is 93,910 km^2 according to FAO's statistics—or with a round figure considering also illegal deforestation—100,000 km^2. It implies a loss of eco-exergy capital of almost 10^{14} GJ/yr. Loss of wetlands by drainage is also significant and could give a loss of eco-exergy in almost the same order of magnitude as the loss of eco-exergy by deforestation. It is therefore important that mankind in the future replace deforestation by plantation of forest and stop the drainage of wetlands to have a continuous increase of the eco-exergy. The two growth forms, growth of network and growth of information, can, however, continue and these two growth forms are far from their limits as discussed in Chapter 2.

The evolution will also in the future follow, according to Monod, a combination of chance and necessity. If a property that is slightly different from organism to organism in a population gives some benefits, an evolution of this property will inevitably take place. Darwin's finches on the Galapagos, for instance, have an evolution towards bigger beaks in dry periods because the seeds are bigger and harder, but only in the dry periods because in the wet periods the beak size is not important or expressed differently as there are no constraints from the hardness or the size of the seeds on the evolution of the beaks. The bipedal evolution of hominids that started about 5 million years ago gave a clear advantage (less energy consumption) in the savanna landscape, which was a result of the climatic changes at that time. The hands could then be applied to other things than to grasp—they could be used to make tools, for instance. To make tools was an intellectual challenge that probably started the evolution of the brain capacity. Tools and bigger brains made it possible to develop the collective hunting of big animals, which implied a more protein-rich food. Proteins are needed for the brain size to grow.

The evolution of the genome is slow. It is hardly possible to see any difference in the genetic information content between *Homo sapiens* today and 30,000 years ago, when he habited the caves in Southern France. It is furthermore difficult to predict in which direction the genomes will develop, because it is hard to see what the next challenges to our survival would be: Resistance to ultraviolet radiation caused by the reduction of the ozone layer? Or social empathy? Or ability to cooperate multidisciplinary, which could mean a development of the cooperative network and the information at the same time?

In addition to the genetic evolution, a cultural evolution that is much faster will take place (Jensen, 2005). If the information technology will continue to develop with the same rate as now, 10 times faster and 10 times more storage capacity per unit of mass for every 3 years, the eco-exergy transfer rates will become comparable with the biological rates in about 30 years. It would require the use of nanotechnology to an extent that we cannot imagine today; but it is of course still an open question to what extent it will be realized. However, it is certain that further development of the society must increase, similarly to the biological evolution, the information content and information rate transfer. The use of material and energy in the ecosystem as well as in the society is quickly reaching the maximum possible, while there are enormous possibilities for the information to continue to grow. We have probably just started to use the possibility with the development of computers and the internet. We are in the very beginning of a new age—The Information Age. In accordance to Moore's law, over the past fifty years, the amount of information that computers can process and the rate at which they process it have doubled every 18 months; the computers today are about 1 billion times faster than 50 years ago. Let us use the Margulis–Levitin theorem to assess the eco-exergy of the ultimate computer. The theorem says that the maximum rate at which a physical system can move from one state to another is proportional to the energy of the system. The more the energy available, the smaller the amount of time required for the electron to go from here to there (Lloyd, 2007). The ultimate amount of energy per gram of matter is stored in the matter and is in accordance with Einstein's famous formula: $E = mc^2$, equal to the c^2. The ultimate computer contains therefore about 10^{11} kJ/g, corresponding to a β-value of 5.1×10^9 or 2 million times the β-value for *Homo sapiens.*

The ultimate computer could register with this amount of energy 10^{31} b/s, assuming the weight of the computer is 1000 g (Lloyd, 2007). According to Moore's law, to reach the level of the ultimate computer would require about 200 years. Of course, we do not know if Moore's law continues to be valid. It would require an enormous creativity and innovative effort. Anyhow, the potential for growth of the cultural information is enormous.

Coda: A holistic and thermodynamic interpretation of the evolution

Coda 1 INTRODUCTION

E.P. Odum's attributes (1969a) give a detailed and well-documented description of the development of ecosystems from the early stage to the mature stage. The attributes are completely covered by the three growth forms (see Section 1.11; Jørgensen et al., 2000; Fath et al., 2004). All the three growth forms imply that eco-exergy stored in the ecosystems and power are increasing, which is not surprising because the two concepts are closely related. Ascendency will increase, too, with some few exceptions (see Chapter 14). The three concepts are therefore appropriate to describe the development of ecosystems—how ecosystems are able to move away from the thermodynamic equilibrium.

The evolution requires of course a much more long-term description than the development of ecosystems. The attributes of E.P. Odum describe the properties of ecosystems that make a development from the early stage to the mature stage of an ecosystem. The attributes of E.P. Odum can be replaced by the three growth forms, as they cover all or most of the attributes. All the three growth forms are accompanied by the increase of eco-exergy, power and ascendency. Similar to the ecosystem development, the evolution can also be interpreted and described by use of eco-exergy and the three growth forms, as it has been discussed in all the three movements of the book.

The evolution has a much larger tool box to make it possible for ecosystems and their biological components on a long-term basis to continuously move away from the thermodynamic equilibrium, that is to gain eco-exergy. The tools or the mechanisms applied to achieve a continuous evolution have been described in Chapters 2 and 3, and it is shown as far it is possible with simple models that they all yield a higher eco-exergy and utilize the three growth forms. The mechanisms are therefore consistent with our thermodynamic description of the evolution by application of eco-exergy and the three growth forms.

As for the development of ecosystems, the growth of biomass or the physical-biological structure has limitations rooted in the conservation of mass and energy, but also dependent on the different possibilities in the different ecosystems, as discussed previously. The limits for the growth of the networks and the information are far from being reached by the evolution up to present. These two growth forms have still (see Chapters 1 and 2) a lot to offer the evolution.

What is the total picture of the evolution from a thermodynamic perspective? How much has the evolution been able to utilize the three growth forms and, particularly of course, the growth of networks and information, because they do not have severe limitations as the first growth form, the growth of biomass or structure? How have the three growth forms worked together to move ecosystems and their living components as far as possible away from the thermodynamic equilibrium by the evolution? We have all the results lined up in the previous chapters and will therefore try to summarize the results in a holistic and thermodynamic description of the evolution in the following sections.

The descriptions of the development and the evolution of ecosystems and their biological components by the three growth forms are closely linked to the tentative thermodynamic law presented in Section 1.11. Let us repeat the support for the tentative thermodynamic law of ecology:

1. The three growth forms covering E.P. Odum's attributes are able to describe ecosystem development and evolution.
2. All the three growth forms imply increased eco-exergy.
3. All the evolutionary mechanisms are accompanied with an increase of eco-exergy.
4. The descriptions of the development and the evolution are facilitated by the use of the tentative law. The results presented in the three movements by the three growth forms and by use of thermodynamics are giving supporting feedbacks to the tentative laws.
5. The three growth forms, the description of ecosystem development and evolution and the tentative law denoted ELT are forming a harmonic pattern that invites falsification and further support.

Coda 2 THE EVOLUTION OF BIOMASS AND THE AMOUNT OF ENERGY CAPTURED BY THE ECOSYSTEMS

We do not know the development of the biomass and the ability of ecosystems to capture solar energy. Let us—very tentative—assume that ponds represent a typical ecosystem 525 million years ago. In that case, the amount of solar energy captured should correspond to about 6000 g detritus/m^2 yr (see Chapter 12), which means 112,200 kJ/m^2 yr. The lagoon may correspond to the typical ecosystems 450 million years ago. It means that the solar radiation captured would correspond to 900 g detritus/m^2 yr or 168,300 kJ/m^2 yr. Let us—of course, again very tentative—presume that the typical ecosystems 330 million years ago could, with respect to the amount of solar energy it could capture, correspond to a coral reef or 27 g detritus/m^2 yr or 504,900 kJ/m^2 yr. Finally, we can be more certain if we presume that the typical ecosystem 35 million years ago would correspond to a typical rain forest today. It is able to capture 70% of the incoming solar radiation (see Table 1.5, Section 1.11). The solar radiation corresponds to about 2 GJ/m^2 yr, 70% of which will therefore correspond to 1,400,000 kJ/m^2 yr. The typical ecosystem today is therefore about 12.5 times better to capture the solar radiation than the ecosystem 525 million years ago. We are moving on thin ice and it is a guess or maybe at least a qualified guess. If the guess is correct, the energy ascendency of the ecosystems has increased 1.5 times 12.5 (see Table 15.1, Chapter 15). It means 19 times since the beginning of the Cambrian Explosion; and the amount of biomass sustained by a typical ecosystem could have increased 1.2 times 12.5 (see Table 15.2, Chapter 15), about 15 times in the same period.

Coda 3 THE EVOLUTION OF INFORMATION STORED IN THE MOST ADVANCED ORGANISMS AND IN THE ECOSYSTEMS

Let us summarize the evolution of the β-values taken from the Second Movement and the evolution of eco-exergy-based ascendency and the eco-exergy sustained by ecosystems from Chapter 15. Let us also find the average β-value for the ecosystems as a function of time by calculating the ratio of the eco-exergy sustained by the network and the biomass (expressed in energy units) sustained by the network. Table Coda 1 summarizes these four sets of values versus "million years ago". The last line "factor" indicates the ratio between the last and the first value of the column.

It is remarkable that the columns 2, 3 and 5 all indicate a factor about 400. The information embodied in the most advanced organisms, the eco-exergy sustained by the ecosystems and the eco-exergy ascendency have all increased about 400 times from 3800 million years ago to 35 million years ago. Due to the increase in the network complexity and organization, and since it is rooted in the increase in the biodiversity, it has been possible for the entire ecosystem to increase its content/capital of eco-exergy (work capacity) 400 times. It means that the eco-exergy has increased by the same factor for the most advanced organisms due to a corresponding increase in the embodied information. The average β-value of the ecosystems has "only" increased by a factor of about 250. The biomass sustained by the ecosystem is, however (see Table 15.2), increased by about 50% (a factor 1.5), which means that during the evolution the information embodied in the

Table Coda 1 Evolution of information of the most advanced organisms and the ecosystems

10^6 years ago	β most advanced organisms	Eco-exergy sustained by ecological network	β average ecosystems	Eco-exergy ascendency
3,800	5.0	220	2.2	52.7
3,200	8.5	288	2.9	57.7
2,300	20.0	763	7.6	96.7
1,500	39.0	2,045	15.7	161.5
750	98.0	3,050	25.4	495
525	133	3,092	25.8	519
480	232	5,218	41.7	675
450	330	9,677	73.3	1,555
330	688	16,412	122	3,510
150	833	38,430	318	7,606
100	980	44,090	322	12,174
35	2127	84,737	547	20,845
Factor	425	385	249	396

organisms has increased by a factor $250 \times 1.5 = 375$. If we presume—see Section Coda 2—that the ability of ecosystems to capture sunlight has increased 15 times since the Cambrian Explosion and probably much more since the first primitive cell emerged, the eco-exergy of a typical ecosystem has increased many thousand times during the last 500 million years. The best estimation would be 6000 times $= 400 \times 15$.

Coda 4 THE EVOLUTION OF THE ECO-EXERGY FLOW RATE

The eco-exergy embodied in the organisms has increased 400 times, but the ability of the organisms to utilize eco-exergy has increased from a rate (see Second Movement) at about 0.1 J/s kg 3800 years ago to 43,000 J/s kg today or rather 100,000 years ago—a factor 430,000! The eco-exergy flow represents the rate at which organisms can *influence and utilize* their environment. The factor, 430,000 times, for the eco-exergy flow density can be considered a holistic conclusion of our application of a thermodynamic angle in our effort to interpret and understand the evolution. The factor 430,000 is rooted in an increase in the eco-exergy of the biomass (about 6000 times, namely 400 times of Kullbach's measure of information and 15 times of the biomass itself) and thereby the amount of solar energy that the ecosystems can capture, an increase in the information embodied in the most advanced organisms about 400 times and an increase in species richness many times (see Chapter 13), which has made it possible to develop more and more complex and more and more effective networks. The three growth forms have worked together to make the evolution towards a higher and higher eco-exergy level of the organisms and of the ecosystems possible.

Coda 5 VERTICAL AND HORIZONTAL EVOLUTION

The scope of this book is to quantify the evolution by a translation of Darwin's theory to thermodynamics. The expectation was that a thermodynamic interpretation of the evolution would give us a new perception and new images of the evolution. The expectations have been fulfilled as it can be seen from the conclusions in the Sections Coda 1–Coda 4.

In addition to the thermodynamic conclusive picture of the evolution drawn in Sections Coda 1–Coda 4, it would be interesting to see to what extent the vertical and the horizontal evolution are coupled and work together. Figures Coda 1 and Coda 2 have plotted the vertical evolution versus the horizontal evolution. The vertical evolution is expressed by the eco-exergy of the most advanced organisms in both figures, while the horizontal evolution is expressed as eco-exergy ascendency in Figure Coda 1 and as the total eco-exergy sustained by the ecological networks in Figure Coda 2.

Both figures show that the vertical and the horizontal evolution have followed the same trends, because the two plots are approximately linear. There is, however, a tendency to a faster vertical evolution up to 330 million years ago, particularly after the Cambrian Explosion, followed by a faster horizontal evolution up to about 100 million years ago. A possible explanation could be that a high species richness has to be evolved before the organisms can be used as components in a more effective ecological network.

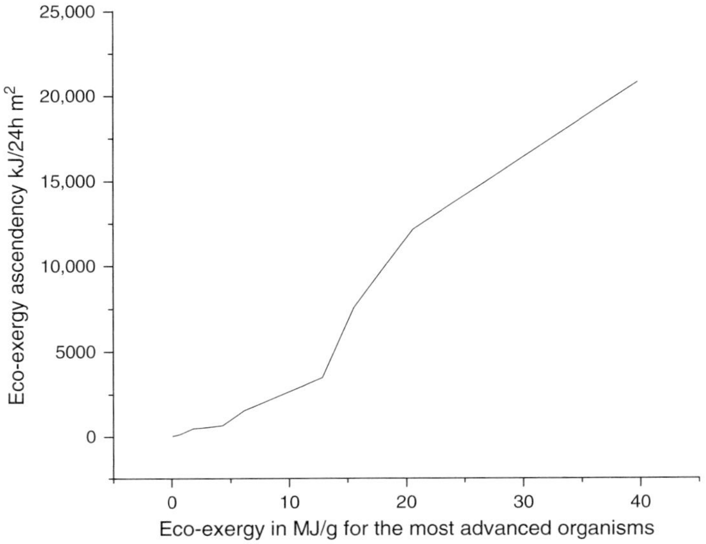

Figure Coda 1 The vertical evolution as eco-exergy in MJ/g in the most advanced organisms is plotted versus the horizontal evolution as eco-exergy ascendency in kJ/m^2.

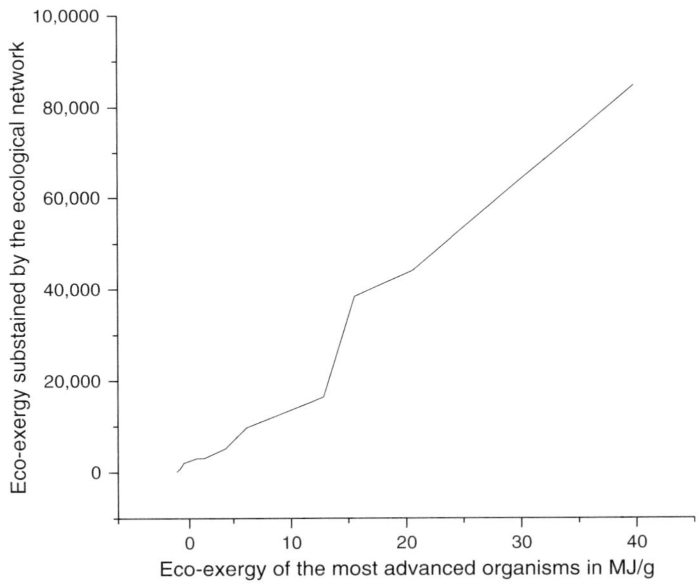

Figure Coda 2 The vertical evolution as eco-exergy in MJ/g in the most advanced organisms is plotted versus the horizontal evolution as eco-exergy sustained by the ecological networks in kJ based on a solar radiation input of 10 kJ.

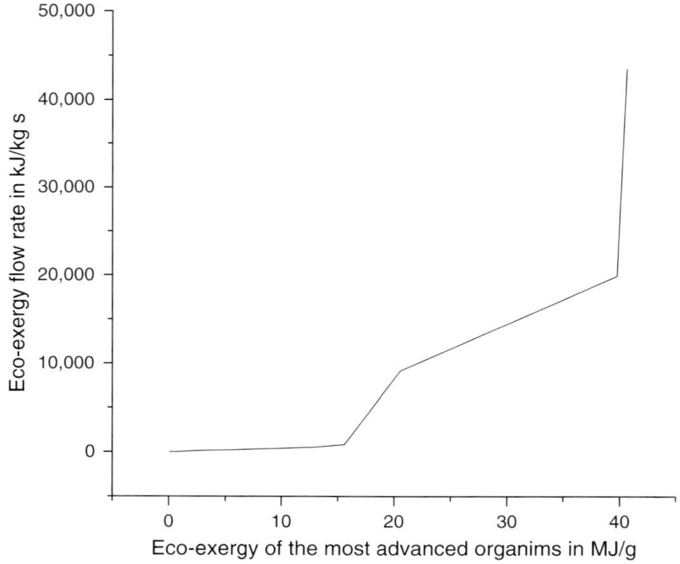

Figure Coda 3 The eco-exergy of the most advanced organisms is plotted versus the eco-exergy flow rate. The rapid increase of the flow rate, particularly after the Cambrian Explosion and the last about 65 million years, must be due to a synergistic effect of all the factors promoting the evolution.

Figure Coda 3 shows a similar plot for the correlation between eco-exergy of the most advanced organisms and the eco-exergy flow rate as a result of the evolution.

Clearly, the plot in Figure Coda 3 indicates an exponential increase of the eco-exergy flow rate relative to the eco-exergy of the most advanced organisms. The increase of the eco-exergy flow rate has been particularly rapid as a result of the Cambrian Explosion and since the birds and the mammals became dominant. The vertical evolution, the horizontal evolution, the increase in captured solar radiation, the increase in information and the ability of the most advanced organisms and the network synergism have worked together to yield a very high eco-exergy flow rate—or it may be denoted eco-exergy power. The result is an enormous increase in the ability of the organisms and the entire ecological network to utilize the environment. The result shows the importance of both the vertical and the horizontal evolution and all the three growth forms. All the factors promoting the evolution have worked together to yield this enormous increase in the utilization of the environment by the organisms and the entire ecological network.

References

Aoki, I. 1998. Entropy and exergy in the development of living systems: A case study of lake-ecosystems. J. Physical Soc. Japan 67: 2132–2139.

Aoki, I. 2006. Min-Max Principle of entropy production with time in aquatic communities. Ecol. Complexity 3: 56–63.

Barbieri, M. 2003. The Organic Codes. Cambridge University Press, Cambridge. 301 pp.

Barrow, J.D. 2003. The Constants of Nature. Vintage, London. 352 pp.

Bell, J. 2006. The Red Planet's watery past. Sci. Am. 295(December): 40–47.

Benton A. 1993. Fossil 2 http://palaeo.gly.bris.ac.uk/frwhole/FR2intro.html

Boltzmann, L. 1905. The Second Law of Thermodynamics (Populare Schriften. Essay No.3 (Address to Imperial Academy of Science in 1886)). Reprinted in English in: Theoretical Physics and Philosophical Problems, Selected Writings of L. Boltzmann. D. Riedel, Dordrecht.

Brillouin, L. 1962. Science and Information Theory. 2nd edition. Academic Press, New York.

Bryson, B. 2003. A short history of nearly everything. Doubleday, UK. 656 pp.

Burnham, R., Dyer A., and Kampe, J. 2005. Astronomy – the definite guide. In Weldon Owen Pty Limited. Sidney, Australia. 433 pp.

Caldwell, R. 2006. What is dark matter and how is it affecting the universe. Sci. Am. 246 (November): 80–88.

Capek, V. and Sheehan, D.P. 2005. Challenges to the second law of thermodynamics: Theory and experiment. Fundamental Theories of Physics, vol. 146. Springer, NewYork. 355 pp.

Childress, J.J., Felbeck, H., and Somero, G.N. 1987. Symbiosis in the Deep Sea, Sci. Am. 277 (May): 39–48.

Christensen, V. 1995. Ecosystem maturity—towards quantification. Ecol. Modell. 77: 3–32.

Christensen, V. and Pauly, D. 1993. Trophic Models of Aquatic Ecosystems. ICLARM and Danida. 390 pp.

Coffaro, G., Bocci, M., and Bendoricchio, G. 1997. Structural dynamic application to space variability of primary producers in shallow marine water. Ecol. Modell. 102: 97–114.

Colaco, A., Deharis, F., and Desbruyeres, D. 2002. Nutritional relations of deep-sea hydrothermal fields at the Mid-Atlantic Ridge, a stable isotope approach. Deep Sea Res. 49: 395–412.

Colaco, A., Desbruyeres, D., and Guezennec, J. 2007. Polar lipid fatty acids as indicators of the throphic associations in a deep-sea vent system community. Mar. Ecol. 28: 15–24.

Cowen, R. 2005. History of Life. Blackwell Publishing, Oxford, UK. 324 pp.

Davies, P. 2007. Are Aliens among us? Sci. Am. 297(December): 36–43.

Debeljak, M. 2001. Application of Exergy degradation and Exergy storage as indicator for the development of managed and virgin forest. (Ph.D. thesis at Ljubliana University).

De Duve, C. 2002. Life Evolving. Oxford University Press, Oxford. 342 pp.

De Wit Rutger. 2005. Do all ecosystems maximize their distance with respect to thermodynamic equilibrium? A comment on the "Ecological Law of Thermodynamics" (ELT), proposed by Sven Erik Jørgensen. Sci. Mar. 69: 427–434.

Dirzo, R. and Mendoza, E. 2008. Biodiversity. In: S.E. Jørgensen and B. Fath (eds.), Entry 460 in Encyclopedia in Ecology. Elsevier, Amsterdam and Oxford.

Douthitt, B. 2006. Saturn. Natl. Geogr. 15: 16–29.

Ebeling, W., Engel, A., and Feistel, R. 1990. Physik der Evolutionsprozesse. Akademie Verlag, Berlin. 374 pp.

Eigen, M., with Winkler-Oswatitsch, R. 1992. Steps Toward Life: A Perspective on Evolution. Oxford University Press, Oxford UK.

FAO. 2005. State of the World's Forests. FAO, Rome.

Fath B.D. and Patten B.C. 2001. A progressive definition of network aggradation. In: S. Ulgiati, M.T. Brown, M. Giampietro, R.A. Herendeen, and K. Mayumi (eds.), 2nd International Workshop on Advances in Energy Studies: Exploring Supplies, Constraints and Strategies. Porto Venere, Italy, 23–27 May 2000. pp. 551–562.

Fath, B.D., Jørgensen, S.E., Patten, B.C., and Straškaba, M. 2004. Ecosystem growth and development. BioSystem 77: 213–228.

Fox, S.W. and Dose, K. 1972. Molecular Evolution and the Origin of Life. W.H. Freeman. San Francisco, California.

Futuyma, D.J. 1986. Evolutionary Biology. 2nd edition. Sinauer Associates. Inc. Sunderland, Massachusetts, US. 600 pp.

Gause, G.F. 1934. The struggle for Existence. Williams and Wilkins, Baltimore. 163 pp.

Geigy. 1990. Scientific Tables. K. Diem and C. Lentner. J.R. Geigy (eds.), 9th edition. Basel, Switzerland. 840 pp.

Givnish, T.J. and Vermelj, G.J. 1976. Sizes and shapes of liana leaves. Am. Nat. 110: 743–778.

Glansdorff, P. and Prigogine, I. 1971. Thermodynamics of Structure, Stability and Fluctuations. Wiley Interscience Publishers, New York.

Grant, P.R. 1986. Ecology and Evolution of Darwin's Finches. Reprinted in 1999. Princeton University Press, New Jersey. 492 pp.

Greene, B. 2000. The Elegant Universe. Vintage Books, New York. 448 pp.

Hastie, N. 2001. Perspective, p. 121–128 In: K.E. Chapman and S.J. Higgins (eds.), Essays in Biochemistry. Portland Press Eynsham, UK. 131 pp.

Haugaard Nielsen, R. 1999. The Archaean Sea were teeming of life. Ingeniøren. Naturvidenskab. Number 6, February 1999, pp. 16–17.

Haugaard Nielsen, R. 2001. Hunting the proteins (In Danish Jagten på proteinerne). Ingeniøren. 16, 20. April: 20–21.

Hawking, S. 2001. The Universe in a nutshell. The Book Laboratory TM Inc., New York. 224 pp.

Heinrich, B. 1987. Thermoregulation in winter moths. Sci. Am. 277 (March): 73–83.

Heslop-Harrison, Y. 1978. Carnivorous plants. Sci. Am. 268 (February): 92–107.

Hirschfelder, J.O., Curtiss, C.F., and Bird, R.B. 1954. Molecular Theory of Gases and Liquids. John Wiley and Sons, New York. 631 pp.

Holling, C.S. 1986. The resilience of terrestrial ecosystems: Local surprise and global change. In: W.C. Clark and R.E. Munn (eds.), Sustainable Development of the Biosphere. Cambridge University Press, Cambridge. pp. 292–317.

Horn, M.H. and Gibson R.N. 1988. Intertidal fishes. Sci. Am. (January).

Jablonka, E. and Lamb, M.J. 2006. Evolution in Four Dimensions. MIT Press, Boston. 462 pp.

Jakobsen, R.K. 2005. We were almost extinguished, when we started to make drawings. Politiken Videnskab og Debat. Sunday the 6th of November. p. 1–2.

Jensen Peter, K.A. 2004. Da mennesket blev menneske. Gyldendal, Copenhagen. pp. 334.

Jensen Peter, K.A. 2005. Human Origin and Evolution. (In Danish: Menneskets Oprindelse og udvikling) 3rd edition. Gyldendal. 432 pp.

Jørgensen, S.E. 1986. Structural dynamic model. Ecol. Modell. 31: 1–9.

Jørgensen, S.E. 1988. Use of models as experimental tools to show that structural changes are accompanied by increased exergy. Ecol. Modell. 41: 117–126.

Jørgensen, S.E. 1990. Ecosystem theory, ecological buffer capacity, uncertainty and complexity. Ecol. Modell. 52: 125–133.

Jørgensen, S.E. 1992a. Parameters, Ecological constraints and exergy. Ecol. Modell. 62: 163–170.

Jørgensen, S.E. 1992b. Development of models able to account for changes in species composition. Ecol. Modell. 62: 195–208.

Jørgensen, S.E. 1995a. Exergy and ecological buffer capacities as measures of the ecosystem health. Ecosyst. Health 1: 150–160.

Jørgensen, S.E. 1995b. The growth rate of zooplankton at the edge of chaos: Ecological models. J. Theo. Biol. 175: 13–21.

Jørgensen, S.E. 2002. Integration of Ecosystem Theories: A Pattern. 3rd edition. Kluwer Academic Publisher, Dordrecht. 428 pp. (1st edition 1992, 2nd edition 1997).

Jørgensen, S.E. 2006. Eco-exergy as Sustainability. WIT, Southampton. 206 pp.

Jørgensen, S.E. 2007a. Description of aquatic ecosystem's development by eco-exergy and exergy destruction. Ecol. Modell. 204: 22–28.

Jørgensen, S.E. 2007b. Evolution and Exergy. Ecol. Modell. 203: 490–494.

Jørgensen, S.E. and de Bernardi R. 1997. The application of a model with dynamic structure to simulate the effect of mass fish mortality on zooplankton structure in Lago di Annone. Hydrobiologia 356: 87–96.

Jørgensen S.E. and de Bernardi R. 1998. The Use of structural dynamic models to explain successes and failures of biomanipulation. Hydrobiologia 359: 1–12.

Jørgensen, S.E., Chon T.-S., Cho W.S. (submitted to Ecological Modelling) Thermodynamic Interpretation and Modelling of conjugation.

Jørgensen, S.E., Chon T.-S., and Rechnagel, F. (in print) Handbook of Ecological Modelling and Informatics. WIT. Southampton. 560 pp.

Jørgensen, S.E. and Fath, B.D. 2004. Modelling the selective adaptation of Darwin's Finches. Ecol. Modell. 176: 409–418.

Jørgensen, S.E. and Fath, B.D. 2006. Examination of ecological networks. Ecol. Modell. 196: 283–288.

Jørgensen, S.E., Fath, B.D., Bastiononi, S., Marques, M., Müller, F., Nielsen, S.N., Patten, B.C., Tiezzi, E., and Ulanowicz, R. 2007. A new ecology. Systems Perspectives. Elsevier, Amsterdam. 275 pp.

Jørgensen, S.E., Ladegaard, N., Debeljak, M., and Marques, J.C. 2005. Calculations of exergy for organisms. Ecol. Modell. 185: 165–175.

Jørgensen, S.E. and Mejer, H.F. 1977. Ecological buffer capacity. Ecol. Modell. 3: 39–61.

Jørgensen, S.E. and Mejer, H.F. 1979. A holistic approach to ecological modelling. Ecol. Modell. 7: 169–189.

Jørgensen, S.E., Marques, J., Nielsen, S.N. 2002. Structural changes in an estuary, described by models and using exergy as orientor. Ecol. Modell. 158: 233–240.

Jørgensen, S.E., Mejer H., Nors Nielsen S., and Teuber J. 1998. The Evolution of the Thermodynamic Equilibrium in the Expanding Universe. Physica Scripta 58: 543–544.

Jørgensen, S.E., Nielsen, S.N., Mejer, H. 1995. Emergy, environ, exergy and ecological modelling. Ecol. Modell. 77: 99–109.

Jørgensen, S.E., Odum, H.T., and Brown, M.T. 2004. Emergy and exergy stored in genetic information. Ecol. Modell. 178: 11–16.

Jørgensen, S.E. and Padisák, J. 1996. Does the intermediate disturbance hypothesis comply with thermodynamics? Hydrobiologia 323: 9–21.

Jørgensen, S.E., Patten, B.C., and Straskraba, M. 1999. Ecosystem emerging: 3. Openness. Ecol. Modell. 117: 41–64.

Jørgensen, S.-E., Patten, B.C., and Straškraba, M. 2000. Ecosystems emerging: 4. Growth. Ecol. Modell. 126: 249–284.

Jørgensen, S.E. and Svirezhev, Y. 2004. Toward a Thermodynamic Theory for Ecological Systems. Elsevier, Amsterdam. 366 pp.

Jørgensen, S.E. and Ulanowicz, R. (submitted to Ecological Modelling) Network calculations and Ascendency based on Eco-exergy.

Kay, J. 1984. Self organization in living systems (Thesis). Systems Design Engineering. University of Waterloo, Ontario, Canada.

Kay, J. 1991. A non-equilibrium thermodynamic framework for discussing integrity. Environ. Manage. 15: 483–495.

Kay, J. and Schneider, E.D. 1992. Thermodynamics and measures of ecological integrity. In: Proceedings of Ecological Indicators, Elsevier, Amsterdam. pp. 159–182.

Klipp, E., Herwig, R., Kowald, A., Wierling, C., and Lehrach, H. 2005. Systems biology in practice. Concepts, Implementation and Application. Wiley-VCH Verlag GmbH, Weinheim, Germany. 465 pp.

Layzer, D. 1988. Entropy in an expanding universe. In: H.B.Weber, J.D. Depew, and J.D. Smith (eds.), Entropy, Information and Evolution. MIT Bradford, Boston.

Lewin, R. 2005. Human Evolution. 5th edition. Blackwell Publishing, London. 278 pp.

Li, W.-H. and Grauer, D. 1991. Fundamentals of Molecular Evolution. Sinauer, Sunderland, Massachusetts. 430 pp.

Lloyd. S. 2007. Programming the Universe. Vintge Books, New York. 240 pp.

Madsen, J. 2006. The Evolution of Life. (In Danish: Livets udvikling). Gyldendal. 308 pp.

Madsen, J. 2006. Livets Udvikling A–Å (Development of the Life from A to Z). Gyldendal, Copenhagen. 370 pp.

Margulis, L. 1981. Symbiosis in cell evolution. Life and its Environment on the Early Earth. Freemann, San Francisco. 420 pp.

Mauersberger, P. 1983. General principles in deterministic water quality modelling. In: G.T. Orlob (ed.), Mathematical Modeling of Water Quality: Streams, Lakes and Reservoirs (International Series on Applied Systems Analysis, 12). Wiley, New York, pp. 42–115.

Mauersberger, P. 1995. Entropy control of complex ecological processes. In: B.C. Patten and S.E. Jørgensen (eds.), Complex Ecology: The Part-Whole Relation in Ecosystems. Prentice-Hall Englewood Clifs, NJ. pp. 130–165.

May, R.M. 1973. Stability and Complexity in Model Ecosystems. Princeton University Press, Princeton, NJ. 265 pp.

Mayr, E. 2001. What Evolution Is? Basic books, New York. 318 pp.

Monod, J. 1972. Chance and Necessity. Random House, New York. 188 pp.

Morowitz, H.J. 1968. Energy flow in biology. Biological Organisation as a Problem in Thermal Physics. Academic Press, NewYork, US. 179 pp. (See review by H.T. Odum, Science, 164: 683–684 (1969)).

Morowitz, H.J. 1978. Foundations of Bioenergetics. Academic Press, New York, London.

Morowitz, H.J. 1992. Beginnings of Cellular Life. Yale University Press, New Haven and London.

Odum, E.P. 1969a. The strategy of ecosystem development. Science, 164: 262–270.

Odum, E.P. 1969b. Fundamentals of Ecology. 3rd edition. W.B. Saunders Company, Philadephia, London and Toronto. 575 pp.

Odum. E.P. 1971. Fundamentals of Ecology. 3th edition. W.B. Saunders Company, Philadelphia, US. 360 pp.

Odum, H.T. 1983. System Ecology. Wiley Interscience Publishers, New York. 510 pp.

Odum, H.T. 1988. Self-organization, transformity, and information. Science 242: 1132–1139.
Odum, H.T. 1996. Environmental Accounting—Emergy and Decision Making. John Wiley and Sons Inc., New York, US. 370 pp.
Ostwald, W. 1931. Gedanken zur Biosphäre. Wiederabdruck. BSB B. G. Teubner Verlagsgesellschaft, Leipzig, 1978.
Patricio, J., Sala, F., Pardal, M.A., Jørgensen, M.A., and Marques, J.C. 2006. Ecological Indicators performance during a re-colonization field experiment and its compliance with ecosystem theories. Ecol. Indicators 6: 46–59.
Peters, R.H. 1983. The Ecological Implications of Body Size. Cambridge University Press, Cambridge. 329 pp.
Prigogine, I. 1947. L' Etude Thermodynamique des Processus Irreversibles. Desoer, Liege.
Prigogine, I. 1955. Introduction to Thermodynamics of Irreversible Processes. Thournes, Chicago. The second edition by Wiley Interscience Publishers, New York, 1967.
Prigogine, I. 1980. From Being to Becoming: Time and Complexity in the Physical Sciences. Freeman, San Franscisco, California, US. 220 pp.
Puzachenko, Y.G. 2006. A global biological variety and its spatially time changes. In: N.S. Kasimov (eds.), Recent Global Changes of the Natural Environment. Vol. I. Scientific World, Moscow. pp. 306–737.
Rasmussen, I.L. 2005. The Solar System. Politikens Forlag, Copenhagen. 260 pp.
Russel, L.D. and Adebiyi, G.A. 1993. Classical Thermodynamics. Saunders College Publishing, Harcourt Brace Jovanovich College Publishers, Fort Worth—Philadelphia—San Diego—New York—Orlando—Austin—San Antonio—Toronto—Montreal—London—Sydney—Tokyo. 620 pp.
Schlesinger, W.H. 1997. Biogeochemistry. An Analysis of Global Change, 2nd edition. Academic Press, Sand Diego, London, Boston, New York, Sydney, Tokyo, Toronto. 680 pp.
Schrödinger, E. 1944. What is Life? Cambridge University Press, Cambridge. 212 pp.
Sharov, A.A. 2006. Genome Increase as a clock for the origin and evolution of life. Biol. Direct 1–17.
Shieh, J.H. and Fan, L.T. 1982. Estimation of energy (enthalpy) and energy (availability) contents in structurally complicated materials. Energy Resour. 6: 1–46.
Shugart H.H., West DC. 1981. Long-term dynamics of forest ecosystems. Am. Sci. 25: 25–50.
Shugart H.H. 1998. Terrestrial Ecosystems in Changing Environment. Cambridge University Press. New York. N.Y. 534 pp.
Skøt, J. 2005. Images of the First intelligence. Politiken, Videnskab og Debat. 21 August. p. 3.
Stanley, S.M. 1973. An ecological theory for sudden origin of multi-cellular life in the late Precambrian. Proc. Natl. Acad. Sci. 70: 1486–1489.
Stewart, F.J., Newton, I.L.G., and Cavanough C.M. 2005. Trends Microbiol. 13: 439–448.
Straskraba, M., Jørgensen, S.E., and Patten, B.C. 1997. Ecosystem emerging: 2. Dissipation. Ecol. Modell. 96: 221–284.
Svennevig, B. 2003. The Life on the Earth. Ingeniøren, Number 18, pp. 16–17. May 2003.
Svirezhev, Yu.M. 1990. Entropy as a measure of environmental degradation. Proceedings of the International Conference on Contaminated Soils, Karlsruhe, Germany.
Svirezhev, Yu.M. 1998. Thermodynamic orientors: How to use thermodynamic concepts in ecology? In: F. Müller and M. Leupelt (eds.), Eco Targets, Goal Functions and Orientors. Springer, Berlin—Heidelberg. pp. 102–122.
Svirezhev, Yu.M., Steinborn, W. H. and Pomaz, V. 2003. Exergy of solar radiation: global scale. Ecol. Modelling 169: 339–346.
Sweinsdottir, S. 1997. Highest probability of life on the Moon of Jupiter, Europa Ingeniøren, Number 21, May 1997, p. 10–11.

Sykes, W., 2001. Seven daughters of Eve. W.W. Norton and Company, London, 306 pp.

Szargut, J., Morris, D.R., and Steward, F.R. 1988. Exergy Analysis of Thermal, Chemical and Metallurgical Processes. Hemisphere Publishing Corporation , New York, Washington, Philadelphia, London. Springer-Verlag, Berlin, Heidelberg, New York, London, Paris, Tokyo. 312 pp.

Thoma, J. 1977. Energy, Entropy and Information. IIASA. Laxenburg. Austria, Number 77–32.

Tiezzi, E. 2003. The Essence of Time (125 pp) and The End of Time (202 pp). WIT Press, Southampton. UK.

Tiezzi, E. 2006. Steps Towards an Evolutionary Physics. WIT Press, Southampton, UK. 158 pp.

Tilman, D. and Downing, J.A. 1994. Biodiversity and stability in grasslands. Nature 367: 3633–3635.

Ulanowicz, R.E. 1986. Growth and development. Ecosystems Phenomenology. Springer-Verlag, New York, Berlin, Heidelberg, Tokyo. 204 pp.

Ulanowicz, R.E. 1997. Ecology, the Ascendent Perspective. Columbia University Press, New York. 201 pp.

Ulanowicz, R.E., Jørgensen, S.E., Fath, B.D. 2006. Exergy, information and aggradation: An ecosystems reconciliation. Ecol. Modell. 198: 520–524.

Vollenweider, R.A. 1975. Input-output models with special references to the phosphorus loading concept in limnology. Schweiz. Z. Hydrol. 37: 53–84.

Von Bloh, W., Fransk, S., Bounama, C., and Schellnhuber H.-J. 2003. Biogenic Enhancement of Weathering and the Stability of the Ecosphere. J. Geomicrobiol. 20: 501–511.

Williams, R.J.P. and J.J.R.F. Da Silvia. 2003. Evolution was chemically constrained. J. Theor. Biol. 220: 323–342.

INDEX